OUR LIVES ARE CONNECTED

Inspired by a True Story

By

STELLA HART

JASON S. BLACKWELL

Booksondemand

Amazon Kindle Edition

Edited by Lara Rosechild

Book cover design by Olivia Pro Designs

Interior book design by Istvan Szabo, Ifj

Printed and published by Books on Demand Philippines Inc.

6/F, Padilla Building, Ortigas Center, 1605 Pasig, Metro Manila, Philippines

ISBN 978-621-8023-29-1

to Joshua Blackwell

SPECIAL THANKS TO

Joseph Denning

Brock Lawrence Sr.

CONTENTS

AUTHOR'S NOTE

True, Science admits that even the unseen things they are delving in cannot be explained scientifically. That is why scientists and philosophers use models or representations. It is to show, to relate by theories, the picture of an invisible world that makes up our physical reality, as well as to unclog the beauty of our world ensnared in amalgamating assumptions. What makes this book unique is the manner of representation employed in order to share our testimony. With the way to portray a twin soul experience as a theoretical model, we are hoisting the wonders of Theology and Physics alike without also confusing the distinction of each one with both.

PART I.
INTRODUCTION

How do you know that the one you've been waiting for is "the one"? How do you know that your hope and faithfulness are based upon truth? Many of us have at least a vague, if not precise idea as to what our ideal "one true love" should look like. Before we can commit and define faith by our preconceived guidelines, we first need to take a hard look at the truth of the process, because it is not often easy, nor clear, to separate the "real-truth" from the "presumed-truth".

If you study eschatology, you will find that followers of every religious group, and their denominations speak of one thought about God, but what is usually considered as faith oftentimes appears to be a legality of culture and instilled traditions. Every group wields unique ways to "recruit" by cutting a clear path to heaven and salvation. What are the "promises" of eternity that kindle every devotee to keep their faith and run the race, and also attract others to join and be a part of the system?

The term "twin souls" is an ethereal concept with varied applications and definitions, however there seems to be at least one commonality to its meaning: Religion is the main course of action when it comes to psychic analysis, while psychology interprets the behaviors and childhood wounds of the people who go through such experience.

Trying to rationalize something "unusual", like a state that you feel for the first time is like teetering on the edge of sanity. You know you are not insane, you are fully conscious, in fact you are becoming conscious more than ever- otherwise you wouldn't inquire.

Over 2,300 years ago, the famous Greek philosopher, Aristotle, might have believed that the creation of both body and soul are like reflections in the prism casted from a single light source. He wrote, "Soul from the beginning is related to its body as a form to matter" [i] Along with the basis of his Hylomorphic Theory, he contended with Plato's Theory of Forms that stated: "Knowledge evolved from souls' experiences in many lifetimes." Aristotle was hence challenged to write a pejorative treatise on philosophy naming it the "First

Science of Theology”, later known as Metaphysics. From Greek words *meta* –which means beyond and the term *physics*, Metaphysics is a multidisciplinary field with main sections Ontology, Natural Theology, Physics, and Math. Ontology is a branch of Metaphysics although Ontology, as a study of nature and existence, is a modern Metaphysics. Specifically, Science and Religion are “partition coefficients” of Metaphysics. The study deals with all that is natural beyond normal. (Metaphysics is often mistaken with Theosophy out of Pythagorean Metaphysics). The intent is to stipulate evidence to the faith and the truth itself, to conceptualize an empirical understanding that advocates that truth.

Let’s say that you are married but at times, you behave as though you are single. Even though at times you accept the truth of your matrimony, you might still feel miserable in the situation. You are not happy because you find little to be happy about, and you cause pain to others, which exacerbates the situation. You often ask, “Why is it that all I feel and see is pain and not happiness? To which should I turn my head to know what it looks like anyway?” Happiness is a state of mind yet to grasp it fully is to excavate the brunt- the reason that makes one relish it.

When your state is somewhat “anonymous” in Psychology and it sounds like a travesty to your religion or it is neither some supernova for the psychics, you might be beyond a trivial state-of-mind. Hence, there are at least three ways to begin a metaphysical study:

- Develop critical thinking and cognitive reasoning
- Understand the philosophy of Math or Physics.
- Relate them all in the Theology

Scientists have worked almost their entire life understanding the complexities of the universe using representational models. But how do they usually prove a theory, without having a concrete answer as to what it is that they are researching in the first place? In this book, we will be “explorers”. We are going to decode the secret behind the algorithm that might take us yonder concerning “souls” through the use of analogy between twin souls and light (quanta). Models or representations as "mental patterns" contribute a preeminent role in describing a phenomenon or unknown by reference to the known and familiar. The important thing will be to put in relation what we don't know with what we do know in order to come up with conclusive results. Models relate a theory using similes and allegories, "but models are independent from theories. After all, models could not perform if they are dependent from theories. According to philosopher Nancy Cartwright, laws and theories can only be described in real world situations by means of models as in terms of analogy." [ii]

Based on actual events, this book considers Metaphysics as an analogical thinking using Modern Science and the Bible in a paradigmatic analysis to illustrate the fundamental theories of light by defining the characteristics of *souls and twin souls*.

"Let us, then, say that there are three forms of contemplative philosophy - Mathematics, Natural Science and Theology... Contemplative study is to be

chosen above all other science, but it is this First Science of Theology (Metaphysics) that we must prefer to all other kinds even of contemplation."

–Aristotle

1
THE BRUNT OF ALL BEGINNINGS

If two people have remained loving each other until they are old, they certainly can be considered soul mates. If we were to ask them their secret, they would likely articulate some typical generalities but they wouldn't be capable of formulating a concrete answer, for the secret lies beyond human understanding, and is still to this day a mystery of the soul. Someone might say that the reason lies in their attraction to one another--that the reason why they are driven to one another is so that they might learn from one another. But the fact remains that such case is yet another mystery of the soul. They found the one, not because he or she was the best one; they found the one because they were perfect for each other, and it's yet another mystery of the soul. That is why they are soul mates.

How can you satisfy or impress someone, or make him attracted to you, or like you? None, there is nothing that you can do. He misses you, he thinks about you always, he wants to be with you, but there was nothing, actually nothing at all that you ever did. You were simply you and he was simply being himself when you were together. So, why the connection? That is the mystery of the soul.

Pure attraction and connection between two people, in a sense, are beyond anyone's control and wishing for a true connection is ultimately futile – it has to be real, from the heart and soul. You will only realize that soul, that you have a soul, when you encounter your counterpart, the mirror of you. In other words, the nature of soul can be realized through an archery of one another called, Relationship: "This is a great mystery, but it is an illustration of the way Christ and the church are one" (Ephesians 5:32).

Nine years, and I'm still in love with the same man who once shared a brief vacation time with me over the course of eleven days. After we parted and endured many years apart, we made plans to be together. Our relationship was fermented through synchronized events, despite 8,000-miles separating us.

There is a two-sided parameter of what love and destiny can do. That it might be possible for a fool to learn the laws of the mysterious universe, but it would be impossible to garner the latent knowledge unless there is epiphany. Past heaped feelings were impossible to release and express in plain narration, but epiphany liberates our cathartic spiritual aesthesia with effortless tolerance that shelters a deeper understanding. It's like faith and love expressed in so many ways but it's the heart, and the spirit that deliberate them with life and truth. It's a *binah* or a processed understanding of one idea from another. Like the discovery of Fermat's Theorem, epiphany occurs not to redefine a meaning but to revalidate the meaning that had been conjectured in the old days. How wonderful it would be to be a part of that renaissance!

Jason said many "one-time" words that replay in my head sometimes; he said them just once yet to this day, they resonate constantly in the haven of a memory. "I just really love you…" "I want to give you a better life..." "I wish it would be me..." But it's not about what I see in him that makes me love him. My amazement is with what God sees about him that I would love such a man. I think God wants all of me, all about me, and this man shows Him the real me.

While I was imprudent, Jason was the type who would never compromise everything just for a feeling. He always wanted all things to be sure- his income, his shelter, his values and how he is able to provide as a man. He's the type who does not say anything or make moves unless there is a deeply-rooted reason. He even changes his mind a lot. Over time, his excuses and staunch standards taught me to act in phase. His prudence taught me to not jump to conclusions only because there were feelings, experiences, opinions, or even "mental inputs". To him, it was all about living in order to benefit events on the long run. After all, the point of all feasibility would be the redeeming purpose of it all. It was not about the short-term – it was the "lifetime consequences" that mattered most – and that is how he understood us and envisioned our lives together… in the long-term.

For the most part, I let him live his own life, we seldom interfered with each other's routines, and we had taciturn months so I asked, "Where is the point of being too much? What is the extreme of being too distant yet dense at the same time, like... the universe?"

The condition was unconventional and ineffable, like

some things can only be expressed in Math. We could not directly discuss the things about each other- he would avoid the topic. We tried to talk it out, but even then, we didn't want to believe.

How will you know whether he or she is your soulmate? It will be transcribed in your first kiss, but even then you wouldn't want to believe. Because if we want to believe in something, it must be that we were able to provide a congruity as to how that disbelief turned into a belief.

I was confident that through this book, Jason's faith would be revived but instead it was I who was restored. Crossroads… Metamorphosis… Polarity... These are all the words that I came up with as I tried to describe the experience, up until the moment I discovered twin soul articles. I was starting to recognize the relationship for what it really was, remembering the things that he had told me.

Sometimes you will only realize that it was all true when the spirit lives on to show you more once the body is gone. Then your frantic mind will say it is too late and your frenetic heart will contend, and then you realize that you have just arrived on time.

As humans, we cannot perceive every detail of the entire idea the moment we cast the first note. Physicist Stephen Hawking's question, "Why are we here?" is comparatively as challenging when we start to envisage our book's objective by taking into consideration how it may run in progress. This is the same way to ask why a certain experience did not come out feeling "full", or sometimes how *answers* may arrive with a *question*.

The latest theory of life origin is postulated by paleontologist Prof. Sankar Chatterjee, who decreed that eons ago there was a massive meteorite bombardment onto the earth's surface. The blast instigated just the right conditions for our planet to be inhabitable – then the first simple organisms soon evolved.

My life began when I met Jason. Before Jason, days were drifting by and time was not fully experienced. They say it's a "big bang", a bolt from the blue that ruptured a moderate yet an unwonted turmoil that shook my apathy. Eventually, everything was ruffled, shuddered, disturbed; at times, paroxysms would stop and there it was again. A certain kind of overwhelming feeling came and went of its own accord when I least expected it, as if the plates of the plane were shifted by the pouring debris here and there.

How will a prosaic soul find contentment or satisfaction if what he/she believes about life is a matter of luck and every event is a product of random coincidences? Some of us may wish to venture the workings of entropy that generated life and habitation because there were chaos, stages, and order, and these juxtaposed polyhedral events certainly buckled us up to look and see, how a seamless merging can bombard and become somebody's universe.

2
RAVEL THE DIVINE

A boy wouldn't go to sleep until his mother would tell him a bedtime story. When she was finished reading the Creation story, the boy asked, "What were those two trees in the garden again?" The mother recalled, "The Tree of Life and the Tree of Knowledge of Good and Evil." The boy rephrased, "If the Tree of Knowledge was from where Adam and Eve took the bad fruit and ate, what was the Tree of Life for?" The mother answered, "If Adam and Eve will to have eaten the Tree of Life, they would have become immortal." After a deep pause, the boy asked once more, "So they weren't immortal in the garden?" The mother answered, "Well, they ate from the Tree of Knowledge. The consequence of which was death. Adam and Eve had everything in the garden and they did not pay heed about those trees. They were so stuffed and so happy that they couldn't understand what death was. Perhaps they thought that death, after such a beautiful life, was as simple as returning to dust and be gone forever."

Midrash, a Jewish rabbinic collection of literature is a traditional interpretation of Creation events, especially the story of Adam and Eve based on the preaching of the Torah. Midrash Aggada explores ethics and values in many biblical texts through storytelling which covers history, biblical commentaries, Philosophy, and Theology. Midrash Halakha attempts to simplify general views and meaning of some biblical texts (REMEZ: deep meaning). Since Midrash searches the entirety of nature and existence based on the pages of Tanakh (Old Testament), the study is often associated with Kabbalah, a Jewish mysticism that probes the parameter of finite and infinite, the relation of the physical and immaterial in our universe qua nature (SOD or secret, mystery). Kabbalah, originated in Chasidic Judaism was believed to be an integral part of the Oral Torah given by God to Moses on Mount Sinai, although there is another theory that Kabbalah began with Adam.

While most people think that Adam was originally created male, Judaism denotes that "man" is not in general "male or patriarchal in this sense, viz". "Adam was created as a single entity in both male and female. In reference to Midrash, Eve was with Adam all along and the two were created together." [iii] This means that Adam could have been created Androgynous, a two-faced man, one of a male, and the other a female as seemingly described in Psalm 139:5: "You hedge me before and behind", depicting a two side of a coin, or the world in upper and lower waters. The Bible text by which the Androgynous Adam was derived from is Genesis 5:2: "Male and female created He them; and blessed them, and called their name Adam, in the day they were created." The conventional belief that Eve was taken from Adam's side comes from the Hebrew word, *tsela*, which means "rib". The Greek Septuagint translated the Hebrew "tsela" to Greek word, *pleura*, which means the "side of the body."

Genesis 1:28 & Genesis 5:2 both describe the 6th day in which mankind was created, and one might consider Adam to mean "mankind". In that day, God commanded man or mankind, "Go and multiply". Based on Special Relativity, one day for God, in His frame of reference upon His command would be many years to effect in our time-frame. God's commanding power to create is infinite but the progression into our world in the duration of time would progress many years for us:

> *"But you must not forget this one thing, dear friends: A day is like a thousand years to the Lord, and a thousand years is like a day."*
>
> *–2 Peter 3:8*

In Athens, between the years 385–370 B.C., a drinking party was reveled in the house of Agathon. It wasn't just any regular gathering-- everyone was engaged in a dithyramb. The topic: Love (Eros). Socrates was on the limelight who at a particular moment asserted that "the highest purpose of love is to become a philosopher or literally, a lover of wisdom." This party had become popularly known as Plato's Symposium from which the term, "platonic love" was derived. Platonic love portrays an original love, a childlike union that draws male and female to spiritual things, like the original, vernal love of Adam and Eve. Today, platonic love is characterized in a relation known as *twin souls*.

Coming from the roots of mysticism, the term “twin soul” was made popular in the ‘80s in the lea of Theosophy. Theosophy is a broader philosophy of Esotericism in which practitioners claim to have the serendipity of the great secrets of the gods and ancient wisdom. Theosophical Mysticism is indebted to Plato while Theosophical Esotericism was commenced with a Russian-German named Helena Blavatsky (1831-1891). An ensemble of “gods” known as Ascended Masters is revered in some theosophy groups of today with names such as Sanat Kumara and Lord Maitreya (whom are also called Christ). The name, St-Germain is common in twin soul writings as an angel or a guide who is believed to be Joseph, the “father” of Jesus incarnated. Some devotees who follow similar groups claim that Jesus is also Maitreya or Buddha incarnated.

Hermeticism, a monotheistic and unitarian theology, advocates that there is a true theology present in all religions. Their creation story is that the original Nous, which is the “Supreme One”, gave birth to another Nous which they call “Word”. The Word created the world and then created Man. The Man then descended to Earth, and fell in love with the physical nature and became one with it. And this is what made man into a dual being: he was both mortal and immortal.

Legend says that twin souls are a single unit that was split and have since then encountered many lifetimes together. It is believed to be the highest form of soul mate relationship. Before oblivion detached them, it is said that they vowed to each other to one day become one once more, as soon as they each had completed a certain mission. This didactic relationship is unpopular to most Christians, but Mysticism fathoms its implication because it is fixated on spiritual arcaneness. However, the panentheistic faith of some religions that practice Mysticism and promote twin souls are not likely to accord with a quintessence that twin soul, being a very exclusive relationship must be an allegory to faithfulness. We may all be the same in this universe but there must be an "only one" to one's life.

Trying to ruminate on a two-faced or Androgynous Adam recounts that Adam was a "peculiar" being, like the distinction of twin souls among all couplings. Adam was in the garden, irrespective of whether it was physical or metaphorical, he was isolated from the things outside of that place or state. Mankind, based on Genesis 1:28, shows two separate genders that might have either chosen their own pair while some could have been pre-matched from the command, "Go and multiply." Or what made Eve different from the rest could be that she was fashioned from Adam's body. In analogy, it portrays the distinction of soul mates, carnal, and twin souls to this day.

I was once puzzled as to how modern-day mystics have come to learn about the epistemics of twin souls. There must be a precursor, an inspiration accumulated in the ancestral school of thought, which might have brought them to this discovery. Jewish Kabbalah was a minor, permitted tradition restricted to an elite few, however, the story of Adam and Eve was an open book question that might have disseminated the idea of Androgynous man (with 4 arms, 2 faces and 2 genitals). The idea insinuates that the dual-gendered human that was "bifurcated" instigates the legendary concept of "twin soul".

From Yiddish (Ashkenazi Jewish), the term *Bashert*

or *Zivug Rishon* which means “meant to be or pre-destined”, is a person who complements you and whom you are the complement to. A zivug is a life partner although in Modern Hebrew, it is termed bat zug, meaning, the significant other. Talmud Ha’rashba in Teshuvos 1:60 relates that Hashem (God) created Adam and Chavah (Eve) together and then separated (split) them. They call this soul splitting, *Plag Nishamasa* or Half Souls- the husband and wife. A popular text from Talmudic Gemara (Sota 2a) states, “Forty days before the formation of a child, a heavenly voice issues forth and proclaims, the daughter of this person is for that person; the house of this person is for that person; the field of this person is for that person.” [iv] This narrative implies that the couple had emerged from the same spiritual root, and marriage substantiates the union not only to build a home, but to “come back home”, which also explains why they leave their father and mother. The rabbis however, are apprehensive in the quest of fatalism as to how the matchmaking splice may be relevant today. A professional matchmaker known as Shadchan hands many criteria before permitting the proposed match to take the lady out to a date. A pre-ordained bashert or zivug rishon is described to be as “difficult” as the splitting of the Red Sea.

Zivug Rishon refers to the first marriage while Zivug Sheni is the “secondary marriage”, a partner whom you personally chose. A Zivug Rishon remains the first, the predestined, regardless of any earthly marriage preceded or followed with a Zivug Sheni. The rabbis enunciated that finding one’s predestined other half is truly a miracle.

A twin soul life is like a narrow road to heaven. Quoting the parable of Jesus, countless of people walk down the same road longing to meet their twin or other half. A lot of them get entangled by sweet talks and propaganda. They were carried off by someone who claimed to be their twin. But a real twin recognizes the other's voice, without the slightest hint of a doubt. There are so many words that can tickle yet his words remain. He said, “I am here, I'm coming for you. Don’t be lost.” As the time trickled by, the road finally got clearer and they were able to meet again.

3
SOUL RECOGNITION

July 27, 2013

"SO, ARE YOU STILL COMING HERE?" Finally, we were able to exchange words over the phone. Throughout the day, I had a feeling that something was terribly wrong, something that had to do with a crucial decision. Would I ever be able to see him again? He asked mc to be strong. I knew that after all, I could not take him away from his family and his country.

"Just go back to where you started," I continued. "I think both of our lives began when you left. Whatever your heart goes through is what I go through…" He could not say a word about our plans, but he let out all of his frustrations. For the second time in many years, we cried together.

May 2008

The chatter named Mr. Silas who had no photo in the

chatroom captured me as I sat in front of my old, bulky computer screen observing the "traffic". Mr. Silas appeared to be the youngest in that woozy chat box, and letters were descending rapidly. You would have to do your best to get anyone's attention by selecting a line to interrupt from, by conforming or contending. Be interesting, be cute, and you're in. This anonymous Mr. Silas was throwing clichés around, which made me wonder who he was talking to. It appeared to me like everybody was ignoring him. The Filipina chatters were more interested in mature western men, in late forties. Not this 24-year old *nobody* whose location was Doha, Qatar. He didn't sound Asian to me. I greeted him in private, which he quickly responded to.

"Your eyes are so kind," he commented on my photo.

"I'm sorry. That's actually my sister's," I chuckled.

We agreed to talk on Skype; when I heard his voice:

"HAHAHA! HAHAHA!" I could not help myself. He got annoyed and I quickly remedied, "Oh, I'm really sorry. I'm just very happy, I don't know why!"

When we opened each other's webcam, I was

surprised to see a very handsome American. Black curly hair, blushing face, and that bashful smirk that gleamed in the darkened room... We fell in in love over our consistent talks.

"Why are you saying that to me?"

"You are more than just a girlfriend. You're my heart."

Our communication over the internet was fiercely occupied. No needle could fall between us. Our messenger was kept open 24 hours. The first thing we usually did by early morning was voice call. Everyday, I'd wait until he'd get off work at around midnight, since we were in two completely different time zones. Although our connection was becoming stronger and fiercer, our topics of conversation never strayed into murkier waters- they were innocent, and nothing could have been purer in content.

After a two-month long-distance affair, our gratification had finally come ashore. I was going to meet Mr. Silas in the flesh, whose real name was Jason. It was July of 2008. Restless and eager, I was hovering at

Mactan Cebu Airport. At nearly 8:30 p.m., Jason had stepped onto the Philippines soil for the first time. At the entrance hall, I was timidly approaching a five-foot, seven inches tall gentleman, whose hazel eyes and lush lashes escorted me towards his stance, then I was prompted by his voice declaring my name in his usual, bashful smirk. About two hours later, we filled the night with kisses and romance, unable to hardly stop for even a breath.

My heart had seen the face of happiness when Jason was in Negros with me. I saw how he turned like a child again when he found a salamander and a *carabao* in the ricefields and those other "peculiar stuff" around. It was a world that he had never experienced before, and he relished in it.

We went on a trip to an island for his 25th birthday. In his scuba diving suit, he kissed me before leaving with the fishing boat. After a pizza dinner together at 9, the place went all dark because of no electricity. We left our cottage to savor the off-shore breeze. The heavens were full of stars and the moon was delightful. I was afraid, but he held my hand as he led me to the water, assuring me that everything was going to be okay, as long as he is there with me. We soaked half our bodies into ocean. The water was cold, but with his arms tightly wrapped around me, I felt safe and warm. We noticed countless glittering objects underneath the surface of the water, and we gleefully started making waves. We realized that the glimmering objects were a tiny school of fish. We agreed that it was one of the most beautiful scenes we had both experienced in our lives. We kissed passionately, and he confided in me then that he could remain in Negros forever.

We took a bike trip in the heights of Don Salvador, in a crisp morning feathered by a gentle wind. The surrounding mountains and green plains were spectacular. On the trek back to the mainland, for 15 kilometers, we were delightfully soaked by the gentle rain.

"No matter what will happen to our relationship, this is impossible to forget," he said, as we spied an amazing waterfall in the distance.

Our last day together…

He chuckled softly as he brushed away the tears streaming down my face, wetting my pillow. “Hey, I will be back,” he whispered.

It was alright to be apart for awhile… at least we would be able to come up with more concrete plans for each other. I knew that we would have a bright future together. Those were the thoughts he left me with as we stood by the gate for the final night in my town.

“You are so lucky. Your boyfriend really loves you,” the driver said as I settled into the backseat of the white van Jason had rented for our route to the airport. He didn’t want me to commute on a passenger bus for a three-hour trip back home, so he had hired the van for me, as well. When he arrived at Cebu, he called me to make sure I was alright.

Four years later, in 2012, my sister owned that exact type of van. For the second time, the van dropped me off at the airport. I stepped out to catch some air from my exasperating trip. Standing in the lobby, I suddenly broke down. Bracing myself, I quieted my emotional outburst and rushed to the ladies’ room, and my tears kept coming.

That was only a flash of Jason. At present, my sister's white van is on my grips…

When Jason left, my longing heart was carried away by feverish insecurities. My time was consumed in waiting for our online chats.

Two weeks after Jason left Negros, at around 2:30 a.m., I sent him a break-up message. He didn't respond and it was killing me. His silence was deep, and it frazzled me with so much guilt that I thought my heart would burst. I kept myself from reaching out the entire day.

That night, he finally responded: "It was so easy for you, like you never cared!"

We argued about how many times I broke up with him. I pleaded him to give me a second chance. He agreed and it was a relief. But things between us changed. He had grown cold. Fearing that I might make another "bad move", I interacted with him as normally as possible as the pain kept on. He started treating me like a friend and there were no more "I love you". Each day, overwrought by tears, I tried to remain stoic. Perhaps he didn't like the way he was acting either, but I knew that if I complained, he would tell me off. I only wanted him to say that he would come back to me, as he said he would. But he gave no hint of it.

I never questioned him about his coldness and eventually, I also stopped saying "I love you". I missed those words and I missed saying them. He eventually stopped our voice calls as well.

Thinking that he obviously need some space, I decided to not go online. Around midnight, his text woke me and excitingly, I logged on to the messenger. After exchanging short greetings, he told me, "You know what, I'm getting tired of you."

Startled, "What did I do?"

PART II.
WHY SOUL IS INTANGIBLE AND COMPLEX

"The more fascinating aspect of the soul concerns dynamics just as the dynamics of physics, the more fascinating part of physical reality... Dynamics is a study of the forces and possibilities generated by the structure." –Rabbi Noson Weisz

4
VIOLETS ARE BLUE, INTENSITY IS TRUE
INTRODUCING QUANTUM

"Our lives are connected…" Jason's voice surprised me over a long distance call at around 11 P.M. in January of the year 2009. But two months before that, a woman had stolen my number from his phone and had called me. The way she introduced herself prompted my recovery, as she brazenly asked me whether I still communicate with Jason. It had been 6 months that we stopped talking, so a female voice coming from his side of the world was somewhat of a bombshell. I never spoke to Jason about the incident until he called two months later. It was easier for me to accept that he had played me for a fool than to believe the ambiguous things he was telling me that night, over the phone. He was telling me about the way he feels, that he still loves me. I informed him that I already have a boyfriend whose name is Johnny.

Johnny was a handsome Irish man. Perhaps some angels knew that as I'd start moving on, I will trip over so there was Johnny. He paved my pain with laughter, he was sweet and thoughtful. But Johnny eventually broke up with me, and all the pain came back, only worse.

When Johnny turned away, Jason was already long gone. After all, it had been almost five months since I had told him to stop contacting me. I wrote to him and several months later, on the first of November 2009, he replied:

"I still love you... You are a blessing in my life."

"Where are you this time?" I inquired.

"I'm back to New Orleans. I lose my job in Qatar."

"Can we skype, please?"

"Promise me you will think that I am with you wherever you go."

Jason had returned to his hometown in New Orleans. Contrary to the luxury he once enjoyed in Qatar, he precipitately incurred crisis working in a warehouse lifting boxes, and he quickly became an alcoholic.

"So, what's your plan for your life?" I asked him in exasperation of his unstable behavior. No reply. Two days later, he notified me that he might be going to rehab. Then our communication stopped for nine months.

After Jason completed his rehabilitation, he enrolled for college through his Veterans subsidy. I was very happy that he made it to total sobriety and started out a healthy lifestyle.

"I am lucky to have courted you." Inside, I began to feel emotional with those words spinning continuously inside my head, so I told him to get some rest since it was late. "I have more confidence in myself now that I am not drinking or anything so I know everything will be better than ever. To live for better reasons than things that sound marketable… I always learn a lot from you. It does make us better. We are blessed."

* * *

What is the nature of light, and why does the Bible

invoke us and call us "light or children of Light"? It seems unimaginable to characterize something which quantities are not matter. Perhaps the only way to grasp is by the embodiment of the physical nature involved. Hence, to describe the characteristics of light is to explore the nature of its interacting bodies, say, the electron and photon. The nucleus allows the atom to function, but without the workings of quarks and gluons, nucleus wouldn't exist. In other words, reality is a product of interactions in the field of undulating particles that generate the things we see and the capacity to see.

> *"What we observe as material bodies and forces are nothing but shapes and variations in the structure of space... That is, quantum waves structures are real and material particles are not."*
>
> *–Erwin Schrodinger*

This scenario leaves us to ask, "What exactly is the law of nature? Is it something "god-given" that sits outside the cosmos and mandates human beings? How significant is it to contemplate whether these laws are descriptive or prescriptive? Amir D. Aczel's book entitled, "Why Science Does Not Disprove God", mentions that electrons exist in a world different from ours, that the universe might be consisted of layers of space or existence, or additional dimensions intertwined into one another. There are several theories that postulate multiverses (or parallel universes) today, as we know, like the String Theory. But for now, Modern Science expands with the Gauge Theory, a type of local theory that designates the parameters of local and non-local, finite and infinite, all classified as Degrees of Freedom. Quantum Field is a framework of Quantum Mechanics that represents its large, infinite number of degrees of freedom, for instance, the shifting of the particles and their world or vacuum. To compare, Classical Field describes electrons in two vectors which are the combined electromagnetism, but if electrons and photons engage in the universal dynamics that navigate the workings of atoms, quantum derivatives are applied:

If the time evolution of the system is deterministic, such a system has six degrees of freedom. If the motion of the particle is constrained to a lower number of dimensions, for example, the particle must move along a wire or on fixed surface, then the system has fewer than six degrees of freedom. fixed surface, then the system has fewer than six degrees of freedom. [v]

Inscribed in the Old Testament, there are two laws the Jews observe: God's Law or the Ten Commandments, and the Mosaic Law or the Law of Moses. If the Jews break any of the commandments, they would offer a ceremonial restitution according to the Law of Moses. However in the first chapter of Isaiah, God showed His wrath over the sacrificial offerings of the people. He called the ceremonial burnt sacrifices detestable because the people's hearts were far from Him and they loved wicked ways.

Some people are zealous in obeying the law. Teachers observe and preach the law; scientists build devices based on the law. But as we incessantly push empowerment into our dejected world, the farther we could seem to get pushed away from the stupendousness of reality. We thrash about negative feelings when our ways to rectify are only palliative to some degree. How do laws serve?

In Matthew 11:25, Jesus expressed thankfulness about the profundity of knowledge that was hidden from people who thought themselves as wise and clever which instead, heavenly things were revealed to the childlike or the lowly. Then He bid everyone to come to Him saying, "those who carry heavy burdens will find rest in the soul". He pronounced this invitation because He was deeply saddened to see His people overburdened with religious and philosophical regulations that were improbable to attain. It was directed to individuals who no longer had the strength or didn't have the wisdom to carry the heavy burden of their personal problems. So Jesus said, "If you will come to Me, your labor will not be heavy and you will find rest in your soul as you will see the reward of your labor." v.28

In John 6, Christ described Himself as the "bread of life". Perplexed, his disciples asked one another, "How can this man give us his flesh to eat? v. 52" Then Jesus explained on verse 63: "The SPIRIT alone gives eternal life. Human effort accomplishes nothing. And the very words I have spoken to you are spirit and life." (flesh=spirit, blood=life / bread of life)

We charge our batteries, we heat food in the microwave, we start a car by pushing a button, but light beyond that scale is the feature that describes the reality of our nature. When electrons were discovered in 1897, scientists realized that Classical Field is short of words to describe energy in the atomic scale. I looked to the fire at my stove boiling for my coffee and said, that fire's hotter, it's blue. Intensity is the technical term for "hot" in Quantum Mechanics. Energy whether in dim or very bright light is fixed but their intensity preemptively differs in terms of frequency. High frequency light has high energy; low frequency light has low energy. This phenomenon intrigued scientists in olden days. Until December 14, 1900, a German physicist, Max Planck, exhibited why energy doesn't grow or diminish, or why energy is fixed and it only differs in terms of frequency. Planck assumed that the "flow" of electromagnetic radiation must come in packets of energy called, *quanta.* As Max Planck and Albert Einstein advanced the notion of quanta, American physical chemist, Gilbert Newton Lewis termed quanta particles as "photons". A single photon is an electromagnetic wave carrying one quantum of energy:

> *Photons, the light particle, are not 'little billiard balls', but a field quanta - necessarily chunked ripples, or "excitations", that 'look like' particles.* [vi]

Back in the 1800s, Phillip Lenard, assistant of Heinrich Hertz compared the very bright light to a dim light. He discovered that the bright light source ejected more electrons than the dim source. This was the first photoelectric experiment. When a coherent light strikes a suitable metal, electrons are knocked out creating an electric current. This emitted electron is known as *photoelectron.* This may sound strange, but a low frequency light source would never free electrons no matter how intense the *heat* might be. *Work Function* refers to the energy required to release the electron from the metal, as the kinetic energy of electron increases with light frequency. This is the work of Photoelectric Effect. Einstein's formula explaining the photoelectric effect is: $K=hv-W$, where 'K' is the kinetic energy of the electron and 'hv' is the energy of the incident photon. Some of its technologies are night-vision camera and the photosensor in elevator doors.

A twin soul is a soul mate relation, but it is the most complex of all since it instigates an actual transformation in energetics. The connection does not submit to the individual's desires, instead they must discover the virtue of the experience. Despite a brief physical meeting, the bond of twin souls is isothermal ($pV=\text{constant}$). Hence, the relationship is more pervasive as it is on "the soul". When things are good, there's bliss; when things are not going well, it's sheer pain. The intensity of this relationship is not found in passionate sex, romance, and sweet promises. Their contention isn't a struggle on how to keep the spark, but on how to keep up with the connection apart. At first you will say, I'm not ready to serve this fervor. Then you will try to balance your state by humbling yourself and limiting your friends.

The merging of twin souls isn't a monochromatic change, like when you have a boyfriend, you have lovelife or when you're married, you have *marriage life*; the meeting invites more provisions in which guidance passes through all areas of their life. Due to their personal issues, however, twin souls can't leap into the reality of love they truly feel for each other. If "moving on" usually means "getting over someone, and start anew", "moving on" for twin souls must be the quantum leap- a sequence in which every leap opens up a new spiritual phase within the relationship.

It is a Quantum Leap when electrons move to a higher energy state as it absorbs the quanta energy of the photon. In an atom, electrons move in angular momentum gradating through a path called, *orbital*. If you live in a big city, you have probably seen an atomic line of light source every night. "Neon" signs are the most colorful. When the electron falls back to a normal orbit, known as ground state, a photon is produced. Photons would arrange themselves in match with the ground state of the electrons in the atom. As a result, they display luminous colors!

Normally, objects around us do not seem to be charged, instead they are electrically neutral hence the term *electrostatics*. Neutral charges contain roughly equal numbers of positive and negative charges. But overall, our electromagnetic field must already be quantized as scientists suggest. Even within the volume of the gases, there is “quantized energy” as particles vibrate and rotate. Truly, connection is about proportions. Energy is basically in-line with the frequency.

From Einstein’s E=hv, energy remains because the intensity of light does not increase with the energy of its photons, but only increases the number of photons emitted. While energy is constant, the transformation from solid to liquid, from liquid to solid formulates work. Transformation is the mechanism, when some of the microscopic system in Classical Physics can be optimized in quantum simulation. Quantization, as a “work function” is called, *Second Quantization*.

Christ had witnessed the traditional forms of worship in his days,that Zion was the only place considered a threshold of worship or people would face the location when they pray, while Samaritans had Mount Gerizim for a place of worship as their ancestors had observed since Jacob. But Christ had said that time will come that location will no longer matter. Anyone can worship anytime, anywhere when true worshippers will worship in spirit and in truth (See John 4:21-23).

Again, light as quantum observes a standard in terms of glow and concentration which must be beyond the comfort zone in order to absorb and emit light and go digital! This itself is the fact that light becomes matter and matter becomes light.

Does God judge based on our church performance? If a husband has to please his wife all the time and he feels the pressure, where is the joy of loving his wife?

The physical cannot control the spirit unless we connect ourselves to the spirit that is higher than us, if our goal is to be a better person. As we can learn from Adam's life, death is the consequence of a knowledge that can't make right- it's called, Sin. Sin does not only separate us from God, it also separates us from a sound mind. Sin is knowing the difference between good and evil and choosing the will of the flesh. Wickedness is choosing evil, and it does not claim sin. Righteousness is a spiritual harmony with God. In a twin soul relationship, there is always a redeeming factor and that is in having someone to reflect with, who understands the root of all the blunders, and such realization is an awakening, a beginning of the most important journey of spiritual retrieval.

During the reign of Herod Antipas, Jesus Christ encompassed the laws of the land with the creed of love calling it the greatest two commandments: "Love God above all and love your neighbor as yourself."

Love is the fire of the authenticity of knowledge and teachings. It is by love that laws are made alive in the spirit. After reading this book, you will learn that the universe is not the one in full control of human activities, or of all creation, but humans, as part of the creation, observe the same laws the universe observes. Knowing those laws is intelligence but observing them is godly epiphany. You are about to learn what those laws are.

5
CRYSTALLINE FISSURES
THE YOUNG'S DOUBLE SLIT

Do souls choose a baby forming inside a mother's womb? Do souls eventually form with the body like how Aristotle exemplified it with wax? He argued, "Just as a wax object consists of wax with a certain shape, so a living organism consists of a body with the property of life, which is its soul." [vii]

There is a very intriguing query raised by the Jewish Talmud that concerns brain transplants: *"What if the head is found in one place and the body in another? And if the body is equidistant from the two cities, from what portion of the body do we measure?"* [viii] This peripheral consideration concerns one's identity as to what happens after his/her brain was replaced of a new brain. Is it still you?

While creationists believe that every soul is directly created by God at the moment of conception, Traducianism implies that only the soul of Adam was created directly by God (perhaps some ancestors too), and the descendants resembled their biological parents, which means that the characteristics of every human being are "fashioned" from their biological family. This was originally developed by Tertullian, a Christian theologian, around 225 AD. It had been widely supported by Protestant and Catholic Churches, while most Evangelical and Reformed theologians are said to be creationists. The problem with creationists' point of view in terms of soul is the question, "Was I created or designed to be poor and unfortunate while others are rich and gifted?"

People are contentiously focused on the context of "Adam the first man" than "Adam the first living soul". 1 Corinthians 15:45 says, "The first man Adam became a LIVING SOUL. The last Adam became a life-giving spirit. However, the spiritual is not first, but the natural; then the spiritual." [ix] When God breathed into Adam's nostrils, He seeded the foundation of life into Adam's mind, and Adam obtained some of God's spiritual characteristics. This, of itself, defines Adam as "the first living man or the father of all the living". He was the first person to acquire God's Spirit, which appointed him as firstborn of all creation in that primordial time. (First man could also mean, "Israel's ancestor or forefather").

The Bible characterizes "living soul" as how a person relates to light. This *quanta* or light in us may be seen in the spirit world but God alone can *configure*. They say it's Consciousness which mostly means, "the higher self" but Job 32:8 simply says, "spirit": "There is a spirit within people, the breath of the Almighty within them that makes them intelligent." The spirit allows the entire soul being to live accordingly; without the spirit, the person is a zombie. The spirit provides endurance to the soul to serve in a higher gauge like the fortitude to do work in gravitational potential energy. As a person, our physical stimuli behave according to what we are as a soul. Since we are not angels, our attributing power is that, we operate in a dual form as both spiritual and physical (matter and non-matter).

A soul in a new-born baby is like an empty disc with instilled features. It's like an idea being brought forth until it grows and matures with the body that will be filled by the spirit through experience and guidance.

While Aristotle did not give a clear picture with his wax analogy how a person got a soul, Apostle Paul described souls with its body as “seed”. In 1 Cor. 14:36-40, he said, “When you put a seed into the ground, it doesn’t grow into a plant unless it dies first. And what you put in the ground is not the plant that will grow, but only a *bare seed* of wheat or whatever you are planting. Then God gives it the new body he wants it to have. A different plant grows from each kind of seed... There are bodies in the heavens and bodies on the earth. The glory of the heavenly bodies is different from the glory of the earthly bodies.”

It is said that twin souls were created as an ovoid of light and was divided into two spheres that share the same “blueprint”. Aristotle described this kind of soul as “a single soul inhabiting two bodies”, which could mean, “two bodies, one spirit”, a spiritual definition of “flesh of my flesh and bones of my bones”. Some writers characterize this domain as *“two souls that happen to share a matching vibration and connect on the same frequency.”* But the fact of the matter is that, the mystery of the qualitative match underlies in the framework of an interacting core.

Scripture does not confirm or deny Traducianism, but it appears that we can't generalize a view on the dispersion of souls, because the way I got my soul could be different from the way you got yours! If you try to compare “souls” to the origination of different types of light, it's quite complex. But this book only focuses on a specific feature of light that may describe soulmates and twin souls.

A famous Double Slit experiment by physicist, Thomas Young (1773-1829) in his Wave Theory of Light exhibits a wave that splits from a coherent light source (laser beam). Basically as light passes through a narrow dual slit, it would split in two then this two would intercept. Passing through the slits, the pair would produce “lines” in a photographic film upon which they strike. Whichever direction they choose produces a shaded band centered on the film. Amazingly, “each photon (or even electron) does not interact with other photons- they interact with themselves, within their own wave packets to produce interference.” [x] So, they split in two and interfere exclusively, and eventually, they return as a single state. This event is known as Quantum Coherence. Coherent light is made up of waves with the same wavelength that are in phase with each other. A great source of coherence is a laser.

Modern setup uses mirrors (beam splitter) instead of slits. It spreads an arriving light beam into two kinds: the reflected beam and the transmitted beam. This incident is mostly viable in making a hologram: "Two waves that are split move from the beam splitter. One gets reflected from a simple mirror while the other gets reflected from the object we want to see on the hologram. They interfere, meaning, they are recombined and they create a system of interference strips on the photographic plate." [xi] Whatever splitting experiment you carry out whether by light wave or particle or whether you use a particle beam or light source, similar interspersion occurs. It shows that Light as both wave and particle is not continuous because its field effect is discrete (quanta).

Waves are usually represented by trigonometric functions. When two pulses come together, the effect is, both waves can be coherent in sine curve and cosine curve that are added to one another. As twin souls coerce equally in the same path, their effects are added together and their depth gets doubled. When it happens, their interference goes on and on and it cannot be cancelled out. Property of wave is $v=f\lambda$ which means speed of the wave equals frequency of the wave times the wavelength. The interaction, when the particle splits and interferes with itself that they tend to appear as one in two states is called, Superposition.

In the attempt to know the position in two states, Austrian physicist Erwin Schrodinger (1887-1961) ran an experiment: the famous Schrodinger's Cat. He set down the cat in a box that contained a radiation detector and hydrocyanic acid to poison it. The purpose was to determine the state of the cat in both life and death (perhaps a near-death experience). What's in between two states was Schrodinger's concern, but why "life and death" to explain Quantum Superposition? The goal was to discern how two states can become a single entity as a result of superposition. The problem was that these two states, *life and death,* and whatever *imago* there is in between were essentially inconceivable. If the intention was to kill the cat, surely the consequence is death, and superposition was hence unidentified. Being "dead" does not hold one value; if death could mean a lot of things, 1 bit (0 or 1) in Quantum Mechanics represents more than one variable, same thing with the state, "alive". In order to follow the details, you must be indwelt in the experience as both "the offender and the victim".

The case of interference can be signified in wave, photoelectricity can be signified in particle, and reflection can be signified in wave-particle; although as a body or system, light is wave-particle.

6
REVOLUTION OF SOULS
PHYSICAL CHEMISTRY

TWIN SOULS are the highest form of “soulmate-ship”, so it is wrong to say that your twin prepares you to a soul mate relationship. Nobody can escape a twin soul bond, he remains in your life especially when you least expect him to; he is beyond the physical... Perhaps one way to subdue this bond is to pursue it with another person who is connected to both of you, somebody who is a part of the syzygy. That person will carry out the dreams, the plans of the lost or deceased twin with you or what you have started. This is the only way, and this may only exist in your fantasy but the picture is not even to break, but it substantiates a tripartite union. It's true though, that the person who has found her twin can easily get connected with people and make friends. It may also be easier for her to find a soulmate, but entering a new relationship is not a guarantee of breaking her bond with her twin. This is not about memories or places that one can just forget and enforce a closure with. It's mostly about the beginning and the fulfillment of the new life since she met her twin. There is a monument, a hallmark of remembrance and gratitude in her entire being.

Soul mates and twin souls were not created because people get married or people fall in love. What is the purpose of the meeting after all? As a representation to one body, we meet our soul mates because our birth or origination is bound to an eternal plan. The paradox is, if twin souls stay close together like normal couples, their eternal purpose is not maximized. Some couples who claim their reunion as twin souls get married and start a family. Maybe that is their purpose, but a twin connection ultimately, is not confined in this lifetime. Allow us to elaborate why.

Scientists and philosophers have been feuding for years as to how knowledge evolves. We learned that Aristotle refused to believe that knowledge augments from soul's past life through reincarnation. Physicist Stephen Hawking believes that human existence doesn't exceed our present life, because the improvement in knowledge and technology would have improved long ago.

Soul mates in eternity come in spiritual imperial family tree, and the progression of knowledge must have flourished from the syzygy or "parallel lives". Transmigration confines the breadth of life because of its worldview that an afterlife is bound on a jaded world, whereas having a soul mate levitates one's horizon by means of hereditable beacon.

After Adam ate the forbidden fruit, God exclaimed, "Look! The man has become like one of us in knowing good and evil. Now, so he won't reach out, also take from the tree of life, eat, and then live forever" (Genesis 3:22). It was a statement that conveys of a consequence, that man had to die for he cannot, with such knowledge assimilated, be an immortal being if he would also eat of the Tree of Life. Thus, the physical body had to die and knowledge had to evolve in our age like a new set of fruits bearing from a distinct tree.

Knowledge through epiphany could have risen from a "harvest" when old "seeds" are replanted and scattered on earth- if you recall the illustration of the Parable of the

Sower. The Bible commonly uses “seeds” to describe “people or group of people”, as what the previous chapter indicates about 1 Corinthians 15. Based on the nature of a seed, the illustration mostly alludes that new seeds were grains sprouted from the old seeds that died or replanted, that imparts a certain legacy. Once this seed dies, a new seed will sprout out of which but this isn't a picture of reincarnation. Although this “new body” could mean “new spirit” which is in Christ, it could also be a literal hierarchy of souls that signifies “one body, many parts”, “a single unit or one nation”, which may relate as epitome of this hierarchy of spiritual families- even soul mates and twin souls. You may want to search “seeds Bible verses”. Here’s another intriguing text from Hebrews 7:10: “For although Levi wasn’t born yet, the seed from which he came was in Abraham’s body…”

> *“The harvest is great but the workers are few. So pray to the Lord who is in charge of the harvest; ask him to send more workers into his fields.”*
>
> *–Matthew 9:37*

A tree is a symbol in the world of atoms, a pattern of multi-group clusters of a molecule. Atom comes from the Greek root word, “atomon” which means “that which cannot be divided.” Plato assumed that water molecule corresponds with the properties of a shape known as Icosahedra. Today, Modern Science exhibits physical evidence to such philosophical presupposition that the symmetry of a water molecule tree or cluster is icosahedral of 14-water-molecule tetrahedron or 4-triangular. These H2O clusters are highly symmetrical to one axis of three-fold symmetry and three axes of two-fold symmetry. It is said that they are particularly favored clusters. The “electronic structure of atoms (ES) consists of 280-water-molecule clusters of icosahedral symmetry.” [xii]

We usually hear that to find a soulmate is to be a good person, to be sweet and thoughtful, to be a hero in every woman's desire. But if one thinks he is good enough, what else is there to challenge about the soul? The truth is, nobody can be sensitive enough to know what areas you will need to work on until your soul mate arrives. No matter how good of a person you are, sometimes you even wonder why nobody acknowledges it. How frustrating it is when somebody close to you doesn't know your innate qualities and inner struggles. Your soul mate does. When, at a certain point you feel content about your well-being, we're sorry to mention that it is rather plausible that nobody will want to be with you because they think you are too refined. Your soul mate is the only person who truly wants to be with you.

Your soul mate can arrive at any moment, but he does not come along to complicate your life or devastate your family. Whatever changes before the meeting will stem from your deliberate, honest decisions. It means that by then you are starting to seek and recognize the truth about you.

Like particles, the search for a soul mate is a mess of hit or miss. We may have no idea how they meet and annihilate in space but a huge "matchmaking device" known as Accelerator shows how they can be harnessed. Accelerators speed up charged particles by large electric fields which attract or repel them. Particles are like a bunch of windsurfers enjoying the ride. Some who slithered faster perhaps thought they are in a competition in finding a pair, but what often takes place is that those in the back are the ones who easily get hitched and paired up and those in the front seem to end up as a bunch of flotsam. By pumping enough energy, these particles get furled by the boost and then the collision occurs. This gaucherie is called an Event. The accelerator hits the target making the two refulgent beams cross one another. This matchmaking is more likely to produce a higher mass particle whose characteristics are determined in a detector.

Standard Model of Particle Physics is about being together or being apart. Tell me who your friends are, and I will tell you whether you're a boson or a fermion. If two particles can occupy the same state, they are bosons. If a particle is a fermion, they don't share the same room or state. Fermions, with members electrons and protons keep a room as far as possible. Although they go hand in hand, they love their own space. Photons and mesons are bosons. They occupy the same room or quantum state, but they do not highly interact, possibly because fermions interact for them, although at most part, two polarized photons are simply oriented perpendicular to one another. Hence, bosons and fermions observe their own intrinsic magnetic moments in spin concurring one caption: "We are too close, but too far apart."

No two particles can occupy the same state, according to Pauli Exclusion Principle- but how come that some particles can be spotted together in one location? Particles of the fermion group have corresponding anti-particle- electron has a positron, and a proton has its anti-proton, and the quark has its anti-quark. They are the same but they are counterparts in two different worlds that's why they don't share the same state. Photons and some neutral particles like neutrino can be seen with their own kind, thus, they do not obey the Pauli Exclusion.

The properties of an atom depend on the order and behavior of electrons as well as the number of protons. Like electrons, atoms follow the rules of the Quantum Numbers n, ℓ, m and s - these are configurations in which atoms decelerate to either lose or gain an electron in order to bond with another atom. We can imagine that orbital serves as pathway designated by quantum numbers. The first quantum number "n" is the energy level. The second quantum number ℓ is the angular, m is thc magnetic quantum number, and "s" is the spin. The values for s are: +½ and -½, also referred to as 'spin up' and 'spin down'. The spin of electrons was first observed in 1921 by Otto Stern and Walther Gerlach.

An unpaired electron will most likely enter an empty orbital rather than entering an occupied one, otherwise, there will be repulsions. When an electron reaches the valence shell, which is the outer orbital, it meets its pair (positron) then they seize these atoms and bond them. Apparently, this phenomenon is called "bonding pairs". Atoms would form a molecule and this molecule will experience a melting point, conductivity, and solubility. When molecules are heated, they bounce into each other and gain energy, to complete these reactions.

While valence electrons have moved to higher orbitals and formed a bonding pair, inner electrons are fledglings that gather enough courage to participate in bonding committals. According to Pauli Exclusion Principle, electrons aren't capable of randomly occupying the highest energy level or orbital where another relationship starts to develop. Also, as electrons move, they make sure that the unoccupied orbital is of the same energy before they can find a pair to fill the orbital with. They can't rush, because there are quantum numbers to observe throughout the process. Moreover, their quantum numbers are unique in configurations and no other electron may stand in their stead.

We can say that valence electrons are the power of the atom- perhaps their most treasured possession. Atoms love to have their outer shells filled with eight electrons, but normally atoms will have to lose an electron or two in order to have its outer shell filled. Only hydrogen and helium seem to be content with two electrons. Therefore, gained or lost electrons result in the formation of atoms. With this inner process in which their neutrons are changing, different versions of atoms are generated- they are called, Isotopes.

The bond of atoms also experience tests, called Molecular Orbital. This operation is like playing cupid and anti-cupid with the help of two electrons in opposite spin that are not destined to match, but simply to test the bond. Their job is to determine and predict the properties of the atom in terms of shape, magnetism, and bond order. The wave functions of cupid electrons interfere constructively, and this leads to bonding while the anti-cupid electrons, classified as "anti-bonding orbitals" weaken the bond of the two nuclei, destabilizing the adhesion of the atom. This test is what keeps the bond from breaking apart. For this particular reason, the number of "anti-bonding orbitals" in a molecule must be less than the number of bonding orbitals. Electronegativity predicts the type of bond of each atom by how strongly they will attract electrons from another atom.

By the use of an X-ray laser at SLAC National Accelerator Laboratory- Stanford, a transition state can be viewed of two atoms forming to become a molecule. Professor Anders Nilsson stated that only a small fraction can go on to form a stable molecule and the rest break apart again. [xiii]

Water or H2O contains an oxygen atom with 6 electrons and two hydrogen atoms with one electron each, which they share with oxygen in the outer shell. H2O is a kind where equal sharing is not practiced. At times, one atom is being taken for granted by the other by spending more time with electrons. And so there are Dipoles, because they have a slightly positive charge on one end and a slightly negative charge on the other. Perhaps the oxygen atom is busy in social life and career, which from time to time causes structural changes in hydrogen atoms. But we know that H20 is a very strong bond. Even if its compound melts or boils, the bond in that molecule remains.

A bond is an exclusive fusion. You can't say, I want to find someone like her man. Think about it, he is a different person when he’s around his friends or his date. The way I am with you might not be the way I am with Jason because he is my closest link.

There will always be someone better, more beautiful, smarter, more loving but if someone says, “I could never find one like you”, it means you are his/her match.

7
LEST WE LILT THE GOLDFOIL
GENERAL RELATIVITY

"... At first, the straight lines which radiate from the starting point diverge farther and farther from one another, but later they approach each other, and finally they run together again at a 'counter-point' to the starting point. Under such conditions they have traversed the whole spherical space."

–Albert Einstein

If we were aware of the distinction between the layers and the boundary, we wouldn't affirm that we are one with all. We wouldn't consider the laws as regulations that impel us to join the majority, but rather we would see reality after reality. Rene Descartes categorized mind and body as one but he characterized both as distinct parts because of the clear perception we obtain through observation.

Western men seem to have a penchant for Filipinas, who they find most attractive, who speak good English, and who revere white men for love and money. Jason was already familiar in Asian trends--he visited Thailand several times before his expedition to the Philippines. As a 25 year-old at that time, his expectation was nothing out of what you would expect from a man his age- drink, party and date. But instead, he met me, and since it was going to be his first time in the Philippines, he made a startling adjustment thinking that Filipinas are like me and not like the Thais. He was a typical nonchalant young man, but Philippines had offered him a lot of new things and pleasures that felt more natural. He was impressed to find that Ilongga women are witty and conservative. It could have only been a memorable vacation for the two of us, but as memories fade, the belief came to light that it was also quite a "meeting". He called my home his home and he called me his heart and we grew together spiritually. He was a city boy whose heart was a sylvan freedom, whose mind was too sharp- he always wanted to be active, not in sports, but in electronics, and to drive in the mountains, learn on his own, explore his potentials and follow his passion, which is the nature itself; a life outside materialism and socialism that put him on restraints. He didn’t like to be told how he must live his life or how he must handle a relationship. He only knew that he must be ready and available for a commitment that he believed would give him a sense of direction and ultimate happiness.

Soul mates and twin souls are different from the usual life partners by choice. With a soul mate, intimacy lasts. You will forget the usual fears, such as the fear of being lonesome. Except for the fact that you miss him, you do not feel empty because he had already completed you. If you agree, Love is the very nature of souls, and with your twin, you will experience everything in the soul; you will know the distinction between the spiritual and the physical self and how both can go together in balance. However, a spiritual connection can be tormenting, not because of what he is as a person or as a soul but because of the physical encumbrance that troubles his soul. You could have wealth, good looks, intelligence, etc., but it's a world upside-down, a world that demands too much but doesn't fill the spirit. It demands because you need the world in a physical degree yet it divides the whole of you.

You will know, ultimately, that he is your twin soul when there is no more room for self-indulgence that you have been crying the blues about like how sorry you are. Those dramas were a blast of ego consuming you, and your twin carries you up from that self and showing you what you really got. You no longer remember your past relationship traumas but of course you don't want another. Your twin had already transported your dramas into his safety lawn, making you totally private and your ego self is replaced of "him" self. It's one reason why he, too, had become vulnerable. Well, it could be a spiritual reunion that stretches at all corners of existence.

Gravity is not a simple verge of attraction where every mass attracts every other mass in the universe, a law expressed by Sir Isaac Newton. In general, that attraction is the density of matter involved, a very exclusive force, a warping curvature of our universe's 4D spacetime, and this, according to Einstein. Our universe has a complicated topology. It changes in shape while maintaining the same local curvature. It is possible, as a matter of fact, that "the density of matter affects the geometry of the universe. If matter is less, the universe is open and infinite. If the density of matter just equals the critical density, the universe is flat, but still presumably infinite." [xiv] Critical Density refers to the average value of matter required at which the Universe is balanced and its expansion has reached its final breadth, after an infinite time.

But if our universe is a closed system, why does it expand? We envision our galaxy like a balloon being engorged and is comparable to a dollop of raisin bread dough. As the bread rises and expands, the raisins move farther away from each other in the dough. Gravitational energy adjusts as distances between stars increase. The galaxies outside our region and our observable universe are in constant movement, as if the infinitude is connected to our observable world as space shifts. Particle physicists proposed 10-dimensional spacetime (or Superstring) alluding that our universe expands to infinity. Based on Hubble's Law, there are galaxies in our universe that we can't see anymore because they have moved away. We think they're gone but they are there; their light has just never reached us. In Thermal Physics, when temperature increases, the particles in the molecule also move farther apart, hence the object expands. This is also true at some range of energy when water is at a freezing point. We call this phenomenon, thermal expansion.

> *"The Kingdom of Heaven is like the yeast a woman used in making bread. Even though she put only a little yeast in three measures of flour, it permeated every part of the dough."*
>
> *–Matthew 13:33*

If the birth of the universe is only by Big Bang or inflation, then there is no Hubble's Law which emphasizes that albeit the infinity, the universe observes a certain bound upon which the whole space and galaxies move.

"The Big Bang is considerably a theory of the birth of the universe that attempts to explain how the universe developed, but technically it does not describe the origin of the universe. According to NASA it is better to think of Big Bang as the simultaneous appearance of space everywhere in the universe… By definition, the universe encompasses all of space and time, so it is beyond the realm of the Big Bang model to postulate what the universe is expanding into." [xv]

"He stretches the northern sky over empty space and hangs the earth on nothing."

–Job 26:7

Critical density. To what extent must a soul grow in terms of knowledge and experience? What is the basis of a person's expanse, or how much matter is needed for a person to reach the equilibrium of success and contentment?

The expansion that meets the critical density brings about a proportion, concordance and satisfaction to the spirit, in accordance to the will of God's infinite wisdom. Like the physical and spiritual, matter and critical density have to meet halfway so that the expansion of a person's life on earth may trigger in an eternal foresight. On the other hand, if the banner of success is mounted heavily upon the material world, the person's universe is finite.

If there is somebody you think of every single day for many years, perhaps the emotional aspect of it is not the only thing that smites you. What if every part of you is emblazoned by his unknowable touch? Galileo Galilei stated that unless a moving body is disturbed, it would run forever in the impulse of haste. Your life was inordinate, it was running straight through a happy-go-lucky deskbound randomness. When you meet your twin, your motion deflects with time. You will encounter many changes, which you might find to be drastic. Light always takes the shortest path, and the straight line is the shortest when the surface is flat (Cartesian). On a curved surface, the shortest path between two points is a curve- known as Geodesic. Spacetime Geodesics moves along the centripetal force of angular momentum. In other words, angular momentum is the analog of linear momentum when applied to a rotational motion such as geodesics. It is a vector quantity, with magnitude and direction and is always conserved. According to Newton's First Law of Motion, all bodies, even the stationary ones, are always in motion, but those that move in a circle experience constant influence of the changing force.

Described by Einstein as a warping curvature of our 4D spacetime, gravity bends the path of light. You are in a relationship that directs your free-fall in a curved line and the relationship itself is the massive body that provides your universe its density.

"What has bending got to do with the speed of light? The quickest path is not always straight."

– Dr. Jack Kruse

We know based on the law of momentum that if everything happened fast due to an instant click, it would be impossible to halt it right away. Angular momentum is constant, and there seems to have no force like an external torque that would stop a conserved system in this direction. Getting into a new relationship just to forget the other person is an extraneous, barricading impulse in which friction is a complete waste to a conserved state. Let nature take its course. If two people move in opposite directions, soon enough they will forget about each other. It's an end-point in "destructive interference", although to achieve such amplitude will take time. If a relationship undergoes superposition, normally, it goes in between constructive and destructive interference so the net result would either be completely constructive or completely destructive. Give it at least one year. If the spark, the thrill, the passion and excitement linger, it mostly means that he is the same way with you. If you live on not noticing that your mind is at ease without him, it means he has forgotten you too. This is how constructive and destructive interference achieve their equilibrium. It's either that the feelings are amplified or that they have simply vanished.

The wonder of existence is that we don't hold the density of this world. Even though we are part of this world, our personal entity is not intertwined to the physical and material one, but to the path. Twin souls basically move in opposite directions, but since their path is close-hauled geodesic, thcy will tend to converge again in the centripetal point.

So if anyone says they can avoid the twin soul connection by switching to another frequency, they aren't grasping the concept of what a twin soul truly is. Twin souls are isolated, and therefore there is conservation in each other's rest mass which feels natural- this can also be referred to the invariant mass. You preserve a certain affinity apart from the throng. Jason's term on this was, "We had a certain discretionary trust".

In Chemistry, "equilibrium" in the world of molecules means "wholeness". A French Chemist, Henri Le Chatelier described the principle saying, if you do anything that messes up the equilibrium of a molecule in the form of entropy, the system will try to be whole again. His exact words were, "A system in equilibrium, when subjected to a stress resulting from a change in temperature, pressure, or concentration, causing the equilibrium to be upset, will adjust its position to relieve the stress and reestablish equilibrium." [xvi]

What makes or determines a closed system is the

constant level that, regardless of the total amount of energy, the state remains where entropy generates work. The creation of Adam accentuates the solitary embodiment of beauty and perfection. Adam needed a helpmate and one reason could be for Adam to get a clearer perception, a distinction of himself, through a counterpart, in the form of a woman. Perhaps his song was, "I've been to paradise, but I've never been to me." Until he met Eve.

When sin separated Adam from God, the composition of harmony or union was shattered in complex interrelations. In order for Adam to represent his identity, he must be detached from himself, to cognize the will of God versus his will. Reunion is putting back the pieces, in a most deliberate reconciliation.

PART III.
LOVE, TWO SOULS IN ONE, MORE COMPLEX

"What if the basic unit was a couple, a relationship, and what if that relationship had an identity?"

–Unknown

8
THE EVER CHANGING OASIS
HEISENBERG'S UNCERTAINTY PRINCIPLE

You might have just made a foolish thing to see how a man must react and respond to your wants if he truly loves you. You'd be privileged to know, if you will know, how he reacts in silence, however, it will take time because he will not show it… if he loves you deep.

If only you could grab him by his collar and force him to speak! But as you look into his eyes you will see love and you will ask, but why am I hurting? And just even before you could look, you will also choose not to speak.

Normally experts say that one indication of a wrong soul tie is when it confuses you. But soul ties are indeed tumultuous. You get confused, obsessed, you're crazy about each other; and of course feel embarrassed about the whole thing. So, twin souls find it hard to know what to do with their exorbitant feelings. Trying to quell it or deny it doesn't change anything. It only aggravates the situation.

If relationships are a configuration space, some laws of Newtonian Physics can advocate the gradient of the dating benchmark. When the road is rough, we adjust the speed. When we are going too fast, we stop accelerating and we maintain a constant speed or we decelerate when pressure takes us to an opposite direction. As you often relate to the classical, what if your own arbitrage of soul mates and love is a predisposed measurement of how much confusions you are to get? Well, this time, it is no longer an account to what is physical!

The fact is, we are used to common measurements in everyday life- meter, ampere, volt, m2... But a soul-based emotion is categorically not "observable". *Observables* are functions on the six-dimensional phase that hold measurable quantity. Observables determine physical output, but a "quantum relationship" is immeasurable as it is cohesive to another soul, and its observable is assigned to probability distributions. Leon Festinger, a social psychologist who postulated Cognitive Dissonance, suggested that the way to deal with the situation or confusion is what counts, rather than reject or avoid it. The way to deal with the uncertainty of

Cognitive Dissonance is when, perhaps, the heart will learn to speak to the mind, “Just do what I feel and I will listen to you.”

It is through the “language of light” that particles communicate, and that is by how they ambiguously give and take (emit and absorb photon). According to the First Law of Thermodynamics, a continual input of energy keeps a system working, hence it cannot expend more energy than it contains and it has to rely on an external source. But what is the standpoint of Quantum Mechanics in terms of energy flow even without an external influence? How will you keep the “keeps” flowing without it driving you mad?

How much love you give is how much love you can keep up with yourself. A starting point at one end can turn into a big force at the other end, and then it could get a proportional amount, like the mutual magnetic field of Mutual Inductance, that is, between the primary and secondary coil of wire of a conductor transferring energies by the number of turns.

The most popular demand, “Love yourself so you can

love others," is not applicable to love itself as illustrated. Moreover, love, based on the principle of Quantum's Linear Operators implies that, two particles must be self-adjoint and they must behave in polarity. The rule is rather perceptible in the verse, "Love God with all your heart and love others as yourself." It means that loving others as much as you love yourself or thinking about them before your own self is the very nature of love.

First Law of Thermodynamics says, "Energy cannot be created nor destroyed" but in Quantum Mechanics and Electrodynamics, "Energy can be spontaneously created and destroyed." Quantum Field relates to the concept that particles like photons and electrons are like bubbles in space that constantly pop in and out of existence. If so, why is energy conserved? Do Classical and Quantum Field oppose one another?

Quarks constantly change their Color as they exchange gluons with one another and convert into new quarks and then more and more energy are added which make the force field stronger. (Color isn't a literal color. It is a mathematical term which could mean state). If a pair of quark moves apart, their color force or interaction increases across the distance. Eventually, they are either pulled back together or they create a quark-antiquark pair. In other words, it is upon collision and scattering that there is an exchange of energy that makes it conserved. The pair could not store all the energy for themselves, into themselves. Quarks are energy carriers. Hence, Quantum Mechanics does not attempt to change the basic laws- in fact it quantifies the constant of the First Law. Thermodynamics does not state how much energy or conserved energy this universe needs but despite the change through annihilation and decay, energy does not increase nor decrease. It's ingenuous from the outside world- it gives life and through which life is taken. That's just how energy works.

The universe is constant, life is constant, but the lives of the people are come-and-go. Love is a stationary field independent of time but the wave function of the lover gets complicated.

When twin souls separate after a short period of interaction, their lives begin as a cycle. Their brief, yet blissful days ended once the clock struck at a particular hour. She left her shoe in his care and their situation apart has become an interchanging of position and momentum.

In a brief lifetime, electron and its counterpart positron (the positive) approach each other, exchange a photon, and then move away from each other. An electron changes its position and momentum when it emits a photon, then the photon would split as a pair of electron and positron. This scattering defines how energy remains. Quantum Electrodynamics simplifies this through Feynman Diagrams that electrons have no fixed direction whenever they emit and absorb photons- they simply change their directions.

"The electron might move to a certain place and time where it absorbs the photon; then move on before emitting another photon at F; then move on to C where it is detected, while the new photon moves on to D." [xvii] This upheaval after the split is that, the position and momentum no longer commute, meaning when you measure the position, you'll get a slight imprecision in the momentum - vice versa. Richard Feynman quantified these visual shorthands as "probability amplitudes". Probability is the square of the total probability amplitude: Probability=$|\Psi|^2$. If the particle spin is directly measured, the act will change or the superposition will collapse and the outcome will be random again. Werner Heisenberg, one of the greatest physicists of our time called this virtual reality, Uncertainty Principle: $\Delta p \Delta x \geq h$.

It is said that uncertainty happens not because our measuring instruments are unreliable, but because there is a whole range of values that the momentum (or position) can take, each with a certain probability. To get the sum of the probability, one has to fire a large amount of quanta. It is mostly convenient to say that a man is a so-called narcissist because it could be that he didn't pass the arbiter's measuring stick, moreover, the grading system has most likely distorted the value. According to the Uncertainty Principle, one of the biggest problems in quantum measurements is the propensity to influence the particles. We learned that observing one particle will tend to affect the other since the act of measurement forces them to relinquish all possible states and will select another location.

We also try to avoid certainty because we tend to change the behavior of the particles when we are looking, so sometimes we choose not to look and assume that we are still right. That's how we mostly coax the resulting behaviour.

According to Irish physicist, George Johnstone Stoney, we must not worry over natural units like inches, metres, or seconds. Rather, we must abide with the so-called constants of nature, for example the gravitational constant and the velocity of light. "If a physical quantity does not depend on unit systems, they are pure dimensionless numbers which values anywhere, and at any time in the universe can be envisaged because they are timeless. Most often, the laws of Physics can be expressed even without a physical constant and that is through non- dimensionalization." [xviii]

9
RUNNER CHASER DANCE
HARMONIC OSCILLATIONS & AHARONOV'S WEAK MEASUREMENT

How does love conquer all and what or who is it contending with? Two ordinary modern-day people experiencing a deep soul connection; they aren't as important as *Romeo and Juliet*, their families and friends do not know one another- but what impedes them from being together? Would King Darius of Ancient Medes say, "The gods keep them apart."? It's onerous not to know who our enemies are: Is it ourselves or the angels who can steer the background space?

The prince turned into a beast, the hero who once saved your life had appeared to be taking it. Or what if it is you who will rescue your prince this time? Perhaps we live by the dynamics of Quantum Physics because when true love does arrive, we might be ruminating with the forces of the universe that it's going to be our self versus the Law.

"When twin souls meet physically and share a few

unforgettable encounters together, they will both feel "drunk on happiness", Sofia Siberia writes. "Shortly after, they will be overwhelmed by the intensity of the connection, not knowing what to do next." [xix] This haywire changeover which they call, "runner-chaser dance" emulates a flash slide of Oscillation, a recurrence of connecting and disassociating. The runner who is mostly the male and the chaser who is the female are perturbed towards a phase transition, as if the meandering part is over and it's time to catch up on a forecast.

I wouldn't say it's complicated; I rather would say that we were trapeze artists in this type of relationship. We trusted each other in the strings of manipulation, which we were not the ones in control of. After a brief romance, we considered each other friends, but at some intermittent occasions, he would declare his feelings and I would say, me too.

When Jason left, I continued meandering, but that was not being free. I was set adrift in what to do with my life and our unwieldy relationship. We ran into each other whenever we had issues with others. It seemed to be a trend in those days, that a new-found guy was the one ditching me with no sign or indication! That was my kind of verticil if Jason and I weren't talking. Each time I had a breakup, Jason would reappear and ask how I was doing, regardless of how long we had stopped talking. Those breakups were like a sign that he was on his way back… or just whenever I would think he's gone.

We were better off alone or alone together just as long as we could share thoughts. We preserved that form of friendship for so long, inadvertently, and probably, for as long as God makes a day all days of my life.

Twin souls are in legal age, single, and love is there, but why can't they be real lovers like others! There is chemistry, connection, they are comfortable with each other, they are best friends, they are happy together! But wait, how has everything turned out perfect at once? A runner would say, "Can we take a slow motion because I can't really classify this, I think it's more of like, you're going to take my life and there's no way I could allow that to happen so I guess I'm running away before we could get more serious, you know!"

A wife or a mother sees her marriage falling to pieces

right in front of her eyes. But her man hasn't simply changed. He was half-hearted, if not unconsciously half-hearted, when he married her. Eventually, he was stifled by the commitment. He used to think or hope that building a family, or to be a good husband and father would resolve all emptiness and impairment. But when passion started drifting away, the man or the woman staggers and then calls it off.

Twin souls tend to disconnect when emotional baggage emerges between them. When the female chaser complains, the runner male annuls himself because it feels like "he can't take it". She doesn't have to nag, because to him, it feels like so much more than just that. Their devotion towards each other wouldn't cover up their issues for them because they would have to resolve whatever emotional baggage first, separately.

The caprice that mainly encumbers the male twin is the discordance between will and action; and the disintegration to show what he truly feels or what he is motivated to do. If he didn't mutually love his twin, (or if he didn't feel the way she does) he wouldn't have such fear. As much as he wants to keep her happy, he doesn't like to open his heart up to her either. As for the woman, a man has to be sure about his plans in life before he can decide whether he wants to commit or not. She wonders what is dreading about her, and why her man wouldn't do so right away, but as for the man, he has to be sure about the relationship before he can make final plans for the future.

People consider marriage to resolve, cover up or fill the pie, a belief system that hangs aloft. If you don't rush to find love, there'll be no more good men left; however, for today's men, it's "man vs. wild". Men falter, playing hide and seek with women who search to "devour".

Twin souls are not dreamy tales made instantly available after a single test. They want to be together for good but what they gain instead is an amplitude of how unprepared they are yet at entering a serious relationship or marriage. Their joy is complete together but with their spiritual smogginess, they feel undeserving and incapable. It could have been easy to be romantic in the soul but romance is not everything about the soul.

In Luke 14:25-33, Jesus said something like, "Are

you sure you want to follow me? Because you may not finish what you started. If you'll build a house, wouldn't you calculate the cost? Otherwise, people will laugh at you if you were not able to finish."

People long for true love but many could not bear true love as it is spiritual. When true love comes, your match will toil that spiritual self!

First of all, what does a man do? He is obsessed with his shiny car or bike, beer with friends; not to mention, he spends most of his time working in the computer and be popular at work. What does a man think? Sports, party, stuff, fast rides, girls. Anything that would distract him or give him deep thoughts is considered "unhealthy". Plus, he loves his space more than anything and he wants to be in full control of everything!

We tried to keep a serious relationship as we were tired of going back and forth and starting all over again and again. Jason too couldn't move forward with his goals. Each time we tried and started, we could not hold each other fast and it's painful to keep repeating especially if it had made us fall in love deeper.

Apostle Paul who made a choice for himself not to marry expressed an admonition: "Do not deprive each other of sexual relations, unless you both agree to refrain from sexual intimacy for a limited time so you can give yourselves more completely to prayer. Afterward, you should come together again so that Satan won't be able to tempt you because of your lack of self-control" (1 Corinthians 7:5).

"Reality" in a more appropriate word in Quantum Mechanics, according to physicist Leonard Susskind is termed as "reproducible". Why do some things repeat over and over again? Quantum Computing, due to probability must be objectively repeated in order to get the sum or a reliable result. Recurrence prepares the state for error-correction algorithms, that whatever phase transitions, accuracy may be achieved and unscathed upon observation.

The purpose of detachment is oscillation and the purpose of oscillation is superposition. Oscillation is a cosine function that repeats itself as it completes a cycle; that cycle is called, Periodic Motion. Throughout the period, there will be confusion and resistivity, then they will individually move ahead with each of their own course termed as, *initial displacement*. Oscillation occurs longer when it serves more inertia, depending on the volume of personal impairment as a person or as a couple, that may affect the sacred bond. Their physical or emotional incapacity basks energy loss that requires a corresponding cycle of oscillation. Hence, this Runner-Chaser is also a Quantum Leap in which every stage is an accelerating force towards the next phase of perturbation. The male twin can't pretend to be ready when he is not-most often he thinks he is but eventually he changes his mind when he senses debility. He cannot believe that he could love that way, that somebody might love him the same.

Periodic motion for twins is an episode of catching up on truth about themselves. They are placed across each other vacillating in a "seesaw", contemplating throughout the disconnection. Speed and time vary in periodic motion that a cycle must be completed, thus, speed has no influence on the amplitude and dilation whether how slow or how fast they must take each other. It is not in their hands to arrange happiness for both since the coupling is arranged in the leading order of perturbations based on the initial condition. In other words, the period of oscillation will depend on how or where one was left in the situation, or a state that is dependent upon the initial condition. She chases after him because she is needy of him but as long as they are not “ready”, they undergo another perturbation.

In 1900s, a group of physicists were running into problems with classical scale for atoms and molecules (except with hydrogen). Although they were comfortable with Schrodinger’s wave formulation, they thought that a new theory is needed to interpret the discrepancy. During those days, Hiesenberg was formulating his Matrix Mechanics that led him to come up with Uncertainty Principle with which he implied that not all forms of measurement must vary in the exact form of certainty. Hiesenberg was right but the impending challenge was on how to overcome the uncertainty. It took almost a century until the year 1988, Jewish Physicist Yakir Aharonov denoted, “Ladies, don't mind your husband too much!” The theory is known today as Weak Measurement.

Today, quantum information has accomplished much

in "seeing without looking". Yakir Aharonov introduced the idea of "Weak Measurement and Weak Values" together with David Albert and Lev Vaidman, published at Tel Aviv University in 1988 with words, "What if we could look a little?" Weak Measurement becomes a preeminent way to "guide" a bond that has turned stable after repeated tests through series of oscillations. Such is a simmering relief to the conundrum in Uncertainty.

> *"In the 1990s, a technique known as 'interaction-free measurement' seemed to promise the ability to 'see without looking... Weak measurement finds what is there without disturbing it."*
>
> *–Aephraim Steinberg, Physicist*

With Weak Measurement, researchers for the first time have efficaciously described how photons move through double slits and form an interference pattern. These photons were emitted by a quantum dot through a double-slit interferometer and by "gently" influencing the photon (sometimes electrons) and by reconstructing its trajectories, scientists were able to come up with some data. It will then possible to factorize large numbers to encode and decode for a more secure information. Scientists performing this are able to predict and "match" the particles by pre- and post- selection in a double-state polarization:

> *Say you want to measure the spin of some particles. You would prepare them in some particular state, say spin up, and you stay focus to that state. This is called pre-selection. Later, you would detect the particles in a final state, and you throw away those that aren't in a desired state. This is called, post-selection. You do this again and again but as Aharonov et al argued, you should measure the spin as gently, as weakly as possible.* [xx]

It is Post-selection when after all the tribulations, twin souls will be left with a number of states that will complement them in a refined self. To observe the relationship's wavefunction is considerably through a “weak perturbative measurement” when they are ready to be influenced “a little”.

Norms think that being touchy will make one fall in love. But spiritual connection is the root of all and the kind of touch in the spirit is unusual than what we normally expect. Actually you don't expect it because the way he talks and laughs about nothing just the mere fact that he laughs because you're there, is already rewarding.

Perhaps what Weak Measurement tries to tell us is, we are to enjoy the experience by studying ourselves instead of claiming it for ourselves. Try to love someone without expecting something in return but for a fact alone that you are both spiritually connected which is the most important thing in life, that is, to keep someone or something that you don't have to maintain, because actually, it will give you surprises that truly last.

Twin souls are simply instruments of a spiritual or heavenly purpose. They are not the ones in maneuver of the connection, which, the purpose is the key to that important task. By nature, a twin soul relationship can get stronger as years go by, and the curse which is, separation, can be a great blessing if God is the center. Since it is a spiritual account, it is only right to enjoy the growth it brings. If you are hurt, it is not always about the other- there is always a room of reflection within; at first, a self-blame being an entry point, but as you move along, your perspective becomes two-way that will expand into an ultimate gratification to God for such gift of bond.

10
YIN AND YANG: THE BINARY OF ONE ANOTHER
DIRAC'S HOLE THEORY

"When meeting your twin soul, you challenge everything you think is you. The polarity itself creates a huge spin to begin the extraction."

–Omnisense

In 1679, Gottfried Leibniz, the father of Calculus was introduced to a binary system of I Ching. His search for a new arithmetic model was taken on an ancient Chinese philosophy: the I Ching or Book of Changes. Leibniz did not consider the I Ching occult; being a Christian, he conferred the sequence pair hexagrams of I Ching with the Christian creed, *creatio ex nihilo* or creation out of nothing.

I Ching is a form of cleromancy or casting of lots which ancient people practiced when consulting a god for a yes or no answer, like the *Urim and Thummim* in

Hebrew Bible. I Ching was elaborated by Tsou Yen (476-221 BC) in a dual state of Yin Yang, a radial symmetrical form that represents male and female dynamics. The man as "Yang" was traditionally remarked in his intellect, in his stiff, self-driven masculinity, while the female "Yin" is soft, receptive and emotional. It is said that twin souls proffer the components of Yin and Yang. For the Jewish rabbis, "the differing feature of Adam and Eve explains the divergent nature of men and women in temperament, intellect, and physical structure. Things like why a woman's voice is clearer and higher than the man, why the man asks the woman in marriage and why he puts a seed on her." [xxi]

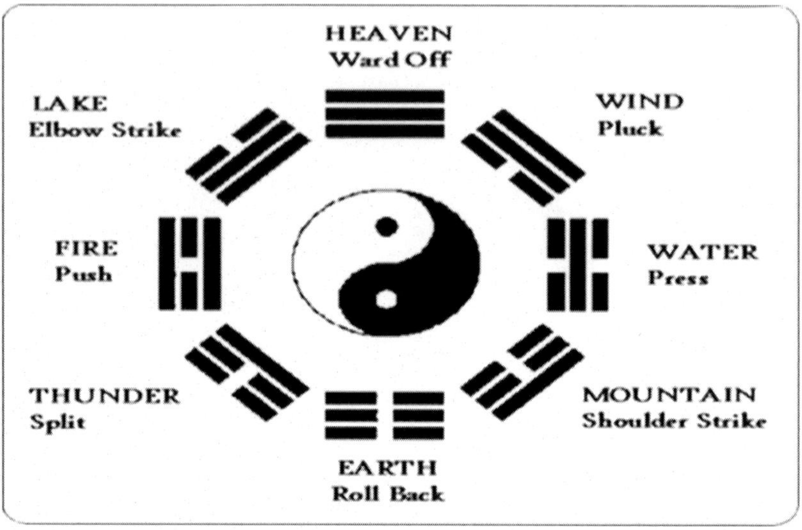

I Ching and Yin Yang

In our first meeting, Jason was the most analytical man I had ever met. He was firm, serious, and stern. I was emotional, dramatic, and dependent. We argued often, mostly on cultural differences. After he left Negros, our ways of argument had turned into inward battles, which always caused profound thinking. Inexplicably not a "couple fight", but somehow a "self-combat". Years passed, the aftermath reversed our personalities. He became emotional while I developed independence. If I was vocal, he was silent.

In 1928, Paul Dirac predicted that in every regular particle, there is an antiparticle partner. It was the first theory to account Einstein's Special Relativity in the context of Quantum Mechanics. In 1932, Philip Warren Anderson discovered the positron as the antiparticle to the electron while the antiparticle to the proton, the antiproton was discovered in 1955 by Emilio Segre and Owen Chamberlain. Dirac Equation relates that a negative electron takes a magnetic spin with its positive pair, the positron. In CP Symmetry (Charge-Parity), this spin moves in opposite direction transforming in left and right, right and left, while maintaining the axes like a mirror reflection; let's say, the right side is the positive and the left is negative.

If the doctrine of the Law of Attraction teaches how to attract the positive, Quantum Mechanics with Dirac Equation and Hole Theory emphasize that both positive and negative must be dealt with. To make a clear distinction is technically an honest way of dealing with both as one, because what takes to attract the full positive is to know its entire value.

Electron is distinctively a left-handed particle but it

doesn't have to be negative forever. As Paul Dirac predicted the existence of antimatter, he asked, "What if we place the electron in a positive-energy eigenstate?" They tried, but this electron decayed into its original negative eigenstate and debased to a lower energy. It shows that the presumed energy state is occupied by another, that in reality, there is one spin-up and one spin-down particle pictured as a vacuum or "sea" called, the Dirac Sea. Those scientists attempted to violate the Pauli Exclusion Principle assuming that an electron may be moved to a positive-energy eigenstate. Hence Dirac thought, "What if this electron has a positive side?" He then introduced the hypothesis known as Hole Theory. The possibility to remove one of the negative energy states from the Dirac Sea is to create a hole in the same mass in the place of electron. This shows the reality of the quantum world, how two or more can be inclusive as one. In Quantum Mechanics, the word, "or" is not "either". Rather, "or" is inclusive in both. Physics Professor Michael Horne classified this strange duality as rebuffing the original definition in favor of the new "both-and" logic, signifying two as a whole, inseparable unit.

You are twin souls because basically your twin is your self, the other part of you, a mirror reflection of you. Your differences only vary on lifestyle, environment, and family life but you and your twin are like hands that need a distinct handedness. Richard Feynman said, "Photons look exactly the same in all respects when they travel backwards in time, so they are their own anti-particles." In polarization, you will find that photons can be right-handed or left-handed depending on which way they rotate.

Why are twin souls a counterpart of each other or why are they totally opposite? That is because they will be merged someday, and if the other is in physical, her potentials will be increased. This will be explained in Chapter 14.

Most often, we can easily detect some people's "true colors" when they devastate us, but we wonder why these people can't see those in themselves. It is also an obstructing view that we see other people's flaws and we don't see ours. That is why the Bible tells us to first remove the log out of our own eyes, and this because our flaws obliterate our vision. What we don't see within is how we behave, and what we see within is what we are. If we are opaque or rough, nothing can be reflected, and so it's hard to tell whether we have truly absorbed light.

Every act of change or transformation is accompanied by pain, but not all pain can create a substance. How do you know that a transformation is converging in quantum, when it does not even allow us to directly measure and observe? If it gives us a rigorous measurement, does that already make it quantum?

Symmetry in Quantum Mechanics expresses a polarity of states of two or more systems that display more planes of transformation while maintaining the subjects' original appearance- they are processed to "look good" together. From the root word, "Syn" which means *together* and "Metry" – *measure*, is the word, Symmetry. Given an infinitely small twist, the result is a consequential invariant of real intimacy between two forms. If one knows the configuration and if the subject is transformable, the artist can go whichever way he wants until it looks exactly as it should in its polarized form.

The more act by rotations, translations and reflections, the more symmetrical the form is.

From –mc2 to mc2, or to become positive is fastidious. It would require an electron to be at least twice its mass times the speed of light. Consequently, what scientists found out was that electron isn’t negative at all- it has a counterpart and this hole is identified as the positron. This hole was an unoccupied eigenstate, which Dirac believed that if the negative energy eigenstates are incomplete, each hole would behave like a positively charged particle. Positive and negative only appear different because they are changeable into one another; in each case, they are coequal where “A + B, has something in common with state A and with state B.” [xxii] They share each other's properties and characteristics, and since measurement cannot be performed simultaneously, this correlation provides a clue about both, through either of them. This intermediary reaction happens if two bodies are part of one system. If the person’s state responds to your state, both of you are correlated. After all, if one cannot be identified through the other, both are not a part of each other. You can't judge a person by your own scale not unless you are his match. You can only assess him by what you are with him.

Since twin souls are close and comfortable with each other, they tend to manipulate each other unconsciously. Yes, there is trust and freedom to function in their individual life but that is simply because of the security level that they feel for having each other, knowing unconsciously that no matter what, he/she will be sticking around. If one gets frustrated, he/she will immediately dwell on the lack and will soon affect the other. For instance, she would say, "I think I will become an old maid," or "I could no longer see places, not until you're here." Or she would criticize his lifestyle by commenting, "That is why you are there (because of the luxury)." And he would say, "There's no hope of finding someone to trust as much as I trust you," or "If only I am there I won't be having problems with…" They get ahold of each other because they will feel extremely guilty if they turn away thinking he might need her and vice versa, and that is because they could sense right away a dual-pain. Thus, they come to a point that romance is no longer important and that what matters is, they have each other.

You can't question this sense of dependence because twin souls are one, literally, but separated as two bodies, not only because they are two different people but their characteristics indicate that there seems to have a huge bulk in between that divides one into two.

We usually hear the phrase, "Till death do us part," but for twin souls the sense of dependence remains, and for the living twin, that dependence remains reciprocal beyond death. Before "a bird" takes him away, she has already secured his heart and his knowledge is sprinkled on her lips. The formerly unwieldy connection is paved through love and epiphany. She lives her life from his breath of inspiration just like when they weren't talking but talking in the heart.

You wouldn't give the Judge a hard time because your twin shows Him what kind of a person you are. He wouldn't have to go through your many activities only to tell you that what you project is not what you are.

How do twins mirror each other? Do they correct each other's faces? The way they judge each other usually starts from baffling questions like, "Why he is that way and why I am this way about each other, and why after all I still love him?"

Your behavior is perfectly reflected back at you, all that you hold inside, including the things you do not even know inside. You will learn to complement his characteristics and realize you just both deserve each other! Sometimes you can't imagine yourself with this person as your intimate partner or spouse but it is this only person you're dying to be with!

The emotional woman will learn through him how to master her emotions. She can't charge him with romantic gestures as indications of love. She gains new experiences with him that show his love for her, like, he loves her stories, he tells her things he doesn't normally say to people, he respects her, he is comfortable and laughs a lot more with her, or the way he reacts to your nonsense and how you utter a phrase that he would exaggerate! The male twin sets his woman apart and treats her different from the rest. She stands out, only because she represents him.

Salvation is like marriage. We don't perform our

duties in order to be saved or in order to “save” the marriage. When our life is committed to the “contract of union”, we keep ourselves pure because we're already in it, because we always remember Him in all that we do.

11
KNOTTY CURIOSITY
QUANTUM ENTANGLEMENT

"...You and I have been struggling with the same loneliness for years. We share the same feelings and we also share our disappointments over the time we have known each other. With you and I, romance or friendship alone, we share the same pain and standards. You always impress me so much. I respect you. I trust you 100%... You are the one woman in my lifetime that I never have to suspect of anything. I say all of this out of respect for a person that I have been blessed to know in my life. I will love you until my death, whether you are friend or girlfriend, or wife. So, I love you."

Jason's mail dazzled my heavy eyes on the morning of January 13, 2013, Sunday. I remember going outside the house laughing in full bliss. After 3 long years, I was hearing from him again say the words, "I love you"!

For the first time in 5 years, I heard his voice once more, and I laughed once again, which already pleased him as it reminded him of the former. Listening to his voice gave me a series of flashbacks. He stayed up all night long telling me that he really loves me and that he is serious about me. He said, he wants us to be together. I told him I want to be his wife.

Saying I love you once or twice in every two years, was to us a courageous act. It's tough when you were also close friends and you talked about everything except feelings. It was tough to love the same person for so long after all the deepest pain and ugly truths, but each time he showed you what was new about him, you fall in love all over again. It wasn't that easy to say the words, because there were solicitudes about whether he still loves you or whether he feels the same as you do now.

Each time we reconnected, Jason and I faced each other as better individuals as though we were each other's quality check. We were seeing a clearer view if we hadn't been talking for months, but of course we couldn't go about that view by simply holding hands for reassurance. He usually kept his messenger off the entire time when we weren't talking. It could go months. Maybe he turned my name invisible, like I usually did. And if we were on speaking terms, he would keep his messenger all day.

When we separated, it was like a whole world split in half and we were on each precipice. There was no bridge to cross over because we shared the same space, and yet we were entrapped. It pained to be connected to him because it got me more in love with him and it was disquieting; it was a giddy sensation which at times made me very happy, and at times was distressing. We were alright when we spoke, but if we became closer, we would separate and not talk for months. On those absences, we were capable of sensing each other- it's the reason why we constantly reconnect.

As I forbore in silence, I never questioned him because I learned to answer those questions for myself. You know the simmering pain in which you feel helpless because you want to be there but you just can't, because you aren't a part of his outside world and so you have no choice but to brood in one corner and wait for the next thing to happen? In these piteous situations, you chase after that slight light. You know you have your own life, but you can't live it out, because the other side of you is laden with poignant murk. He was never vocal of his frustrations- he would ruffle just one word about it. He was the type who would laugh and drink off his barren devastation; he endured them all and appreciated himself whenever he could stand again.

* * *

“Action in a distance” is a mutual, effortless, yet forceful unanimous activity in the quantum field that proves that our space isn’t nothing. We know that in Coulomb’s law, electromagnetic force holds an aptitude to attract and repel charges in which an object involved is moved without the act of touching it. Even repulsions are just amazingly right, because atoms do not fly apart. In a force called the Casimir Effect, two uncharged distant plates facing each other can affect photons. Furthermore, in 1957, physicists David Bohm and Yakir Aharonov discovered that an electron passing outside a cylinder could still feel the effects of the field inside the cylinder.

At times you can never be surc whether what you feel is your own feelings or your twin's feelings even amidst other people. Without the bridge of connection, you will be carefree again thinking that he / she is just one of the many failures. It's like, thinking about someone is different than if he is naturally present in your thoughts.

Non-locality, as physicist John Bell (1928-1990) suggested, is the lifestyle in our microworld system. Bell said that without this spooky action in a distance, there is no Quantum Mechanics. While Superposition allows a particle to exist in the combination of two or more states, Entanglement happens when two particles share the same state despite the distance. When superposition maneuvers interacting bodies, the nature is that of Entanglement.

The common way to produce entangled photons is by the strike of a laser passing through a special crystal- amongst these crystals are lithium iodate and barium borate. Most of the time, a coherent light or quanta would go straight through a non-linear crystal and it's a rare occasion that the beam produces a light particle that splits in two. Their correlation can be predicted in polarization, and when one passes through a vertical polarizing filter, the other will take the exact opposite direction instantaneously. Imagine a vast number of photons coming out from an ultraviolet laser. This pair of photon that splits has combined energies and momenta in equivalent to the original photon from which they emerged. Together they will be measured or "pre-judged" through SPDC (spontaneous down-conversion).

> *"When two systems, of which we know the states by their respective representation enter into a temporary physical interaction due to known forces between them and when after a time of mutual influence the systems separate again, then they can no longer be described as before, viz., by endowing each of them with a representative of its own. I would not call that one but rather the characteristic trait of Quantum Mechanics."*
>
> *–Erwin Schrodinger*

Alice and Bob share an entangled photon pair. Alice, being entangled with Bob, transforms Bob's photon through her photon, for instance, she could disentangle Bob's photon; Bob could as well perform it. How does this situation relate to some people's personal lives?

Some entangled photons of Alice and Bob pertain to their individual social life, family life, religious life, and career. These major areas interrelate in Alice and Bob's lives. Since this couple is entangled, the events serve as signals and people close to them are also reflected. If Bob loses his job, Alice gets one; or if Bob gets promoted in his job, Alice eventually opens a new business. One's event will have an effect on the other, then there is going to be an inverse superposition. (Depending on the position and momentum according to Einstein, the effect to each other is not always simultaneous).

As Einstein described Entanglement as a "spooky action in a distance", he argued that these particles must inherently have possessed a certain spin before it could be measured. These particles, he said, are like a pair of gloves, which means that the phenomenon is rather local and deterministic than it is nonlocal and probabilistic. Einstein was saying that a left hand glove had the right hand glove, since the beginning, and that these particles did not randomly choose a pair to be with because they always had each other, as if they were one, and even apart, they seem to be always together.

In 1950s, David Bohm and Yakir Aharonov used spin as property, instead of *position and momentum.* These particles came from one source that made them entangled. If particle A reveals spin up in the z-direction, particle B, if measured in the z-direction will reveal a "down" spin. To Bell's Inequality, the initial state of a particle before making a choice on which spin to take is not as simple as getting a binary outcome of spin-up spin-down, 0 or 1, or left-right. After all, how can we anticipate an identical opposite outcome? That may be possible in the classical, but in Quantum Mechanics, 0 can be both 1 and 0, and 1 can be both 0 and 1. This is due to the coherence in which superposition is the measuring system. An entangled photon pair experiencing superposition possesses each other's properties, as if they are one but aren't related! Assume a collection of my state and his state. Our two superposed states make 2 bits already, 0 and 1, and our value or sum is our post-selected, polarized state. A "byte" is a collection of 8 bits (e.g. 01100011).

In the late 1960s, after reading Einstein's paper, John Clauser of Lawrence Berkeley National Laboratory constructed a machine to test Quantum Entanglement. This machine could measure thousands of pairs of entangled particles and can compare spins in different directions. Since particles can in a sense be at many places at one time, our computers today can reconnoiter larger solutions in a snap.

Nature, as portrayed in twin soul paradigm, conveys compatibilism that freewill is dependent on the brunt of deterministic causation and remuneration. What if he did not meet me? What if we were planned to meet but we did not?

The Free Will Theorem of John Conway states that a choice among a 3-perpendicular direction is not a predetermined function of the past. I was observing my swiftlet birds flying around from one point to another as I waited anxiously which spot they would choose. The initial state was random but their choice of location must then be a rational state. It's freewill. I set up a spot for them and once more I was hoping they will like it and they did. It’s their choice although it was indeterminate why they liked it. Randomness is often mistaken for free will. But Freewill Theorem says that freewill and randomness are different from each other.

It's very easy for people to say that God is manipulative, it's also easy to say that God neglects people and there is no God. I believe that God had perceived what would happen to Jason but because of freewill, God, in general, does not perform tricks to change an incident especially if a sojourn was a course of path from an inscrutable past. The only indication I can share that God loves Jason, is when He fused south and north so that Jason will meet me and will have my heart. We once lived shattered lives, we were two hard-headed people, but our link had offered a remedy for us to know ourselves or with whom, and with what our spirit must find shelter upon. Two is still better than one because they will give each other a beautiful meaning when everything appears meaningless and unjust. Astronomer John Byl noted, “If God is directly responsible for quantum events, this entails that these things are therefore predictable by God. Hence we arc left with a deterministic universe, at least at the quantum level." [xxiii]

God's light is the amplitude made available to all and those who open their hearts to God will get to absorb and emit light and be perturbed in all significant states until God's purpose is made known.

12
THE INNERLIGHT SINFONIA
RESONANCE AND TELEPATHY

Internet was getting boring and annoying because our spirits longed for each other's touch. The ocean, the stars, the mountains have truly missed the hedges of our sultry glow. Yes, my spirit grieves for people whose cellphones seem to be the only touch they handle. When we're together, his hand, his eyes, his lips are all mine. He is very attractive in his whole self.

In a single sentence and laughter, we have that talk. Our fricatives are utterances of our own language. I miss his bawdy reactions and comments. We don't run out of things to talk about and even if we wouldn't talk for months, some matters would hover over to our next conversation. He is not an artist, but a lot of times his words surprise me. His every sigh is like a spring that enters my heart and stops me from questioning. Or maybe he is an artist, because he can enjoy things with me. We enclosed our hushed moments with music. He would always bring a stereo with us everywhere we went, and a battery spare for that stereo. He would take me outside for a walk or just a sit side by side. He would point his finger to the things he would see whenever we were on the road, and I have become that way.

In our 11 days together, the present or the future was never a worry. We roamed everywhere, and it's the mountains we loved the most. Negros was a paradise and it will always be. His favorite scented candles lit in a *nipa* room and the starry ocean-field were the things we had as luxury. We laugh a lot together but there is profoundness in our personalities that silence us about ourselves.

Telepathy, also known as distance feeling, is believed to be one form of communication for twin souls. Telepathy may be defined as the transfer of emotional or mental energy, or a combination of these, from one living being to another without physical contact. People are getting an impression that twin souls are telepathic because the couple reflects each other's innermost feelings. It is said that the life they steer is projected dually: one with their environment and one with their absent twin. The truth is, their matched frequency is not a form of psychic or telepathy. They are simply inherently resonant.

The case of telepathy and clairvoyance remains a surmise in the religious and the scientific world. However, there is no technology to prove telepathy, if you expect it as a mind medium that transmits an *sms* to another's brain. Two linked people, in a deterministic feat can have more than six degrees of freedom, so they can be empathic. These people are artists. Through music, breeze and visual colors, these empaths are receptive to the sonority of the sunset, mountains, ocean, certain people... Twin souls develop this sensibility after they are physically separated.

"In 1953, German physicist W.O. Schumann of Technical University of Munich discovered that the earth's cavity produces vibrational pulse. Later, by the use of Schumann Resonance, physician, Dr. Anker Mueller discovered that the earth has the same natural resonance as the human brain." [xxiv]

We have learned that true connection has nothing to do with the classical electrical lines. Connection is discrete, it will show you everything. Everything is not just too many, but absolute in all fragments. Connection lets you experience each one, each day of the absolute.

Dr. Wolfgang Ludwig, the father of Pulsed Magnetic Therapy stated, “Manmade frequencies drown out nature…” How could man-made frequencies divert us from natural frequency? Ludwig emphasized that “the prevalent manufacture of man-made radioactive resources like computers, TV's, cell phones, air conditioners, radio waves, and so on, produce damaging electromagnetic fields, which we can‘t see but may feel in the form of stress, headaches, sleeplessness, and moodiness. These frequencies interfere with the body‘s own electromagnetic operating system and a resonance of a healthy, happy life.” [xxv]

The reflection of light comes from the inside, but to infuse light into another in order to induce life, the recipient must have joined the light. This could be a total internal reflection, like a light injected into an optical fiber transported to the other end of the fiber that we commonly use in internet connections. When people are preoccupied but not resonated to what they inherently believe is right, there is internal conflict.

"...Even Gentiles, who do not have God's written law, show that they know his law when they instinctively obey it, even without having heard it. They demonstrate that God's law is written in their hearts, for their own conscience and thoughts either accuse them or tell them they are doing right."

–Romans 2:13-15

Will you fancy seeing a cybernetic feature of yourself standing in front of you? You may find that underneath the lovely smile, there are brindled spots, because some twines could be disconnected while others are tapped. We have a little chandelier in the living room that has three lanterns. Whenever we turn it off, one lantern is left open. The dim looks horrifying.

Light coming down from the sun is white, and this white light strikes in a spectrum of different colors. The wavelength in the range of 400 and 700 nm of radiation is distinguished in varying colors. These wavelengths are frequencies of the white light, but the lengths and how they extend colors aren't visible to the naked eye. When frequency strikes an object, it causes electrons to vibrate in high amplitude, and then the object absorbs the light. Our eyes cannot distinguish between random and polarized light, but to put this in simpler terms, the interacting light is polarized and it gives our sight, the photoreceptor cells of our eyes, the interpretation of what colors we see. This interdependence is a matter of Resonance.

Contrary to the classical belief that the arrangement of molecules is random, newest simulation shows that the arrangement of each alcohol molecule is locally ordered and beautifully designed due to their OH bonds that have the same resonance frequency. This experiment was conducted by Matthijs Panman and his colleagues from University of Amsterdam. If the infrared light that illuminates the alcohol is endowed with the right frequency, some OH bonds in alcohol molecules will vibrate as a result of resonance synchronizing with the light wave.

Based on QED, twin souls share one "loop". This loop is a circular path in which energy transfers take place. Illustrated in the Feynman Diagrams, energy loop represents the directions of photons and electrons and how they emit and absorb photons, in relation to time. It's probability because they are all over the place, and you know it's real because they are complementary and systematized although immeasurable.

As you take a glimpse on infinity, you will try to grasp an incalculable system through Polarization, in which, after a series of "messy" oscillation and superposition, particles are oriented to an axis. A vast number of photons passing through the crystal are "segregated" through the strands of a polaroid filter. The wonder is the way this material prevents some of the waves from passing straight through, while allowing others to pass. Light waves are distributed uniformly in all directions, so only a very small fraction of waves would pass through a polarizing core. Polarization is a wave property that corresponds to particle's spin. The sum or the product will be revealed out of the particles that are able to pass through the polarizing filter. One of the most common products of polarization is the liquid crystal display (LCD). The first polarizing filters were constructed by French scientist, François Arago in 1812.

Anything that works together, and anything that makes up light, energy, and matter is from God; it is reality, and anything that is short to that reality, in scale, in the workings of quanta and intensity, for transformation, is an outcast of reality as a whole. It's true that twin souls give each other a sense of direction, but the reason behind their meeting is to understand how they are "scaled". TEKEL, from the book of Daniel means, "You have not measured on the balances (superposition) and have not measured up (Daniel 5:27)." If a person could not be justified or measured up, he is frivolous, short, and unqualified.

It's true that after death comes the judgment but this judgment has a process for a disembodied soul who is in *shock*. A part of his judgment is his relation to the physical life. If the soul has a twin soul, he will be focused on her living twin and he would care less about the life that he left behind as if he starts living a new life with her. The reason though is not only bodily but more than anything, he relates to her spiritually. While Christians look forward to the Great White Throne

Judgment, some people believe that as long as this Great Judgment has not yet taken place, some souls and some angels hold the prerogative to serve their cause in the physical world. Afterlife is not a hall for family reunions or church assembly. Whether a Christian, a Jew, or a Muslim, he will immediately find the truth about himself and his Creator.

I don't watch TV series but when Jason was in Qatar, I became a fan of a Latino *tele-novela* about a rich man who died and returned as a personal driver to his young wife. The woman married him for his money but she fell in love with the driver who happened to be her deceased husband reincarnated. The moral is, she fell in love with him as a person and not for any material gain. Acceptance is what every soul needs, and it's about knowing yourself to get the same from another.

When, at a certain point in life, you think you have finally accomplished everything- brilliant children, a nice house, a stable business, a fine marriage, and photos of your tours, you say, what a blessed life; you say, you have given all your potentials and you have used God's gifts right. But sometimes as you spend a moment alone, you wonder about yourself with an inexplicable void within.

When your physical world shatters right before your eyes, what's left will be you and the light that you will reveal. The soul must be filled with much light to offer. Has something or somebody helped you ignite lately? Keep the music playing. We can only be prepared by keeping our lamps filled.

PART IV.
SOUL DEPARTS, SPIRIT REMAINS IN LOVE

"... A male-plus-female marriage is a signpost or a signal about the goodness of the original creation and God's intention for the eventual new heavens and new earth."

–N.T. Wright

13
TWILIGHT CRESCENDO
QUANTUM ELECTRODYNAMICS

It's amazing how electromagnetic waves feel free to propagate or radiate even without the continuing influence of the moving charges, however, in the development of QED, Max Born, Werner Heisenberg, Pascual Jordan, and Paul Dirac discovered that in the crusade of perturbations, many integral parts are divergent. A photon after passing through a crystal can split at the same wavelength, each with half energy but some of them differ in wavelengths. This divergence is also true in hydrogen atoms, which scientists classify as Degeneracy. According to Dirac and Schrodinger, states with the same n and j quantum numbers but different l quantum numbers must be degenerate, meaning, they have different quantum states but equal in energy. In 1947, Willis Lamb and Robert Retherford showed that the $^2S^1\!/\!_2$ and $^2P^1\!/\!_2$ states of hydrogen atom were not degenerate- the *s* state had only a slight higher energy. "Lamb Shift (from Willis Lamb) describes the small difference in potential level between two energies, say,

$^{2}S_{1/2}$ and $^{2}P_{1/2}$ of hydrogen atom. This effect provided the stimulus for Renormalization Theory." [xxvi]

I'm having a thought as to what form to use in hydrogen, whether singular or plural but it is plural because there are 2 hydrogen atoms in a single hydrogen molecule. These atoms share similar states but their quantum numbers differ in terms of the magnetic moment of their electrons in different orbitals.

In a hydrogen molecule, hydrogen atoms in an experimental "metal box" experience chaotic electric fields that push the electrons inside them. Of course this causes the electrons to move slightly further away from the proton in the nucleus. As a result, this shift due to the electric fields moving around the hydrogen, lowers the interacting force of the electron to the proton. It's the independent photons within the metal box that are actually responsible for shaking and pushing the electron! Like Quantum Leap, Lamb Shift happens because the electron experiences a peak at the center of the atom. This explains the penumbra in the energy states of two hydrogen atoms (that form a hydrogen molecule). One of the hydrogen states, say, the $^{2}S_{1/2}$ is left with higher energy after an electron was pushed towards the photons, (this Lamb Shift only makes a very slight energy adjustment for the hydrogen atoms).

Let's take an illustration. I had many boyfriends, typical boyfriends. One man came along. His love was infuriating because he was not like the rest of the boyfriends I had known. But my father admonished that if he was like the rest, how could I have understood what I'd been doing, unless I want to repeat the same cycle? This means that real compatibility is about divergences. Two people correlate in a feeling and are settled in the same state in different environments. However, if neither of the two will accelerate, then there is no energy shift. A relationship that experiences an energy shift lifts you up. Due to an external weak magnetic field, two entities are split into "two levels" each, called the *Zeeman Effect*. In other words, the energy splitting m*j*g which differs for each state will become four Zeeman levels.

From an energy shift or Lamb Shift, Feynman Diagrams show varying divergences in the loop in which photons and electrons are oriented by a process known as Renormalization. Renormalization (Normalization) is a procedure that calculates the effects of low-energy photons and electrons how they are to be "qualified" at high energy levels. By renormalization, variations of energy are conflated to equalize the state of the photon and electron in the energy path of the loop. "In Zeeman Effect, when external magnetic field is stronger than the internal magnetic field, the orbital and spin magnetic moments become quantized independently" [xxvii] Ultimately, everything about the hydrogen couple becomes "normal" or arbitrary.

Based on Newton's Inertia, your twin who holds the same mass as you do, caught you off guard, poked your head, and showed you why the same blunders and issues in life do not go away. Then Quantum Electrodynamics would stipulate that the energy of you both will be split, diverged, divided (in quantum states), renormalized, and then normalized.

In contrast to the ultraviolet divergence, the infrared divergence does not require renormalization which we wondered why. Photons and gluons belong to infrared group. If you recall, photons do not obey the Pauli Exclusion, accordingly they are protected by Gauge symmetry from getting a mass. The differencc is that, if people have ego, electrons that exist in different world from photons have ego too- it is categorized as "self-energy" represented by a sigma symbol of Σ. Renormalization resolves the divergence problems for the self-energy electron/positron by arranging its vacuum polarization to behoove with the massless photon. Several forms of symmetry are simply local and this self-energy is oftentimes a result of changes that it itself causes in its environment. When self-energy is required to assemble some distributions of constituent charges from infinity, gauge symmetry comes in through renormalization. So, ultraviolet divergences which include electrons undergo renormalization while infrared divergences which include photons, protected by Gauge Symmetry, do not undergo renormalization. Electrons are considered abelian and it is commutative while photons are non-abelian and non-commutative.

Connection works for two people whose calloused hearts are fragmented through each other to eradicate the ego so their hearts can absorb light. By renormalization, divergences must be proportionate based on Zeeman

Effect.

You will know, ultimately, that he is your twin soul when there is no more room for self-indulgence that you have been crying the blues about like how sorry you are. Those dramas were a blast of ego consuming you, and your twin carries you up from that self, showing you what you really got. You no longer remember your past relationship traumas but of course you don't want another. Your twin had already transported your dramas into his safety lawn, making you totally private and your ego self is replaced of "him" self. It's one reason why he, too, had become vulnerable. Well, it could be a sort of reunion that stretches at all corners of existence.

"Dark Night of the Soul" of the mystics, is a term that relates major spiritual revivalism in which a person is said to be "emptied out". Christians comparably call this state, Repentance. Repentance is not simply a feeling of remorse. It is a deliberate decision to turn away from sin into embracing a life with Christ as the Lord of your life and Savior. Jesus said in Matthew 11:11: "Of all who have ever lived, none is greater than John the Baptist. Yet even the least person in the Kingdom of Heaven is greater than he is!" So, the process of how John would become a heavenly person particularly in the replica of Elijah is a "spiritual" exemplification of "Renormalization". You can imagine that none of us are qualified in God's Kingdom, not even John in his days. But those who are saved are in the process of becoming like our Bridegroom, Jesus.

We can't search the reality of ourselves by searching God through our body or through the universe. We die in ourselves first and be born in spirit. We can't live our entire life spotting each and every BS we got, or chasing after blind spots and trying to ignite them each time, until we get exhausted and end up totally bored and dull. If an outburst had made you feel you're dying, don't be scared, just get up and prepare to meet your new self. That could be the major one-time transformation, and that could be everything you ever needed to shine. It's not magic, and even *accepting Christ as your Savior* is not a magic word that will automatically make you a Christian. Transformation starts from small realizations taken boldly but the upshot must be about dying to self as an ultimate consolidation, leaving the old self behind. It could have taken 46 years to build a temple but with Christ, it would take 3 days to rebuild it after its ruin!

What can we learn from Non-Commutativity of non-abelian groups? Sure we all want to find an ideal partner; we are even predetermined as to what we need to desire about someone. A twin soul relationship shows how we can be worthy of what we desire of. We have an instilled idea of chivalry, and we want to grab that fruit and consume it in an instant, but that is a shortcut to Commutativity. The effect is illusion, in which the experience had never rooted from within. The goal is to make a twin soul relationship commutative, like how much Einstein wished for Quantum Theory, that after tracking down the gradient of probability and non-commutativity, it would be as determinate as Newtonian Mechanics. But most likely it won't happen anytime soon on this earth.

For an incident that is hard to understand, my question was not, "Why did it happen?" My question was, "Why did it have to be so overwhelming?" As a young bloke, Jason did all he wanted to do and traveled a lot. So I wondered what made him stop doing the things that he used to do. He was a succcssful young man, intelligent. He could still drink and travel if he wanted to, nobody could stop him. He would be angry if anyone gets in his way. But that's not what happened to him. He chose to seek peace and true happiness. He managed to throw away his bottles and sat down and made plans with me.

A twin soul relationship is convoluted in the process, but whatever the situation, a state can be renormalizable when divergences are working out the way it should be. And that was when he transformed his twin.

Stop rationalizing that your emotion is the only thing that makes you human. Inspired by Feynman's words, think that you are renormalizable and your vacuum of space is polarizable.

14
TRANSLATE SERENDIPITY
QUANTUM TELEPORTATION

"Someday, after mastering the winds, the waves, the tides and gravity, we shall harness for God the energies of love, and then, for a second time in the history of the world, man will have discovered fire."

–Pierre Teilhard de Chardin

"A body that has lost its soul is no longer potentially alive," said Aristotle. "But if a living thing's matter is its body, then that body should be potentially alive by definition." Jesus said, "What do you benefit if you gain the whole world but lose your own soul? Is there anything worth more than your soul?" Even though the words mostly mean, "What is earthly wealth if you will lose life?", they can also be signified for soul mates as, "What is earthly life without you? I may have all the riches of the world but without you, everything is meaningless."

“Can’t live without you” or “Die without you” are heart-rending lyrics that seem to be too much for euphuism. You might find the line ridiculous, denoting a suicidal despair because of a loss but the truth is, those words are just fit and nobody is truly gone. This phrase of itself has biblical and scientific significance. What was your life before Christ? It’s random, shallow, and your regular days were simply a bunch of mess cobbled together. It felt right, it felt bad at times- you’re enjoying the ebb and flow of dissatisfaction. This is also true in the life of electrons. In QED, electrons wobble, they do anything they like until they find a life-changing experience, a *vertex function* in which they would annihilate with another and get entangled!

> *“The electron does anything it likes. It just goes in any direction at any speed however it likes, and then you add up the amplitudes and it gives you the wavefunction.”*
>
> *–Richard Feynman*

If G-d will send His love particle, the connection, the entanglement, will transform the pair. This act is to destroy the old and recover the original. If a machine scans me, it will reveal a great deal of information that relates to my partner's information. The "teleporter" sends my data to him and vice versa, to reconstruct us. Since measurement distorts the quantum state, my old self must be destroyed and be replaced with a new me. My twin and I will be a replica of each other. The feature will be that I become like him and together we are a new person... as one.

In the island of La Palma in the Canary Islands, a state-of-the-art lab performs Quantum Entanglement with Teleportation led by Austrian physicist, Anton Zeilinger. One entangled photon stays in La Palma while the other is sent by laser to the island of Tenerife, 89 miles away. Then Zeilinger sends away a third photon to interact with the entangled photon in La Palma. As a "middleman", the third photon correlates the quantum states of the entangled pair and translates the information of both. Eventually both are transformed into an identical copy of that third photon, the intercessor.

A sarcastic post on the internet said, “If a man speaks in the forest and there is no woman there to hear, is he still wrong?” EPR Paradox suggests that by observing the position of the woman in the city, we gain the position of the man in the forest. A link that connects them must provide us the probability of the entire system. Two particles linked together must be a representation of the other so that the sum of all information can be acquired from either of them and neither one is disfigured. The solution is found to be a redeeming factor which itself transforms the entangled photon pair as identical to each other. This is the purpose of Teleportation. In cryptosystem, it is one way to protect the information, so that others could not intercept and decrypt the information transmitted.

Teleportation is like a twin prime conjecture in which the pairing of prime numbers is mediated by a composite number. Let’s say, my particle Anxious X sends a signal about how much she cares for Anxious -X located thousands of miles away. Of course I also want to know if the man handling Anxious -X particle cares for me. The problem is, with such anticipation, my measurement will disturb the state of my Anxious X particle because even though I care, my judgment may distort my Anxious X. This means that confusion affects judgment because my Anxious X particle is yet in the verge of superposition, and measuring it will force Anxious X to release the state of uncertainty. This situation is referred to as, *Projection Postulate* in which the particle is projected into one of the states in superposition. It will be difficult for me to come up with a sound decision, or I would struggle which one could be the most appropriate excuse to consider- well, it is yet in superposition!

Quantum Criticality is a newly discovered technique of transformation among condensed matter in the properties of solid or liquid. Pioneered by Lev Landau in the 1940s, Quantum Criticality shows that condensed matter emerges an "order parameter" (of order and disorder) which is one sort of symmetry breaking called, Quantum Critical Point (QCP). Quantum Criticality is interestingly used for levitating superconductors. Superconductors are materials in which electrons pair up and travel without friction, so no energy was lost in the journey. In Superconductivity, electrons are conditioned in between states of insulation and conduction which means their density is placed at a higher degree of freedom. It is said that this newly discovered phase transition is not a result of thermal fluctuations, but by quantum mechanical fluctuations as light is indeed a discrete quanta.

> *Quantum fluctuations are responsible for the profound transformations in electronic properties, found to develop near a quantum phase transition. This property makes it possible to magnetically levitate superconductors.* [xxviii]

A fake, besotted heart is too fragile- it can't perceive the reality of the soul, but as you submerge into the spiritual self, you will find the standards of the flesh unappealing. Twin souls connect because their souls are no longer fused in the world- but in each other's frequency. Most areas of their lives are charged to each other; those areas are vestibules that fleck them to be sonorous to each other through remembering and contemplating. As they move away from ego and romanticized love, they become more spiritually connected. This hereof indicates how a soul mate bond gets stronger. A spiritual heart is the only thing that can offer unconditional love. If we are one with the world, and this world will sink, we will sink too. What is density and what is the purpose of superconductivity after all? Subserviently, what is the use of being connected with my partner if the connection itself will eventually end in destructive blankness?

> *"When two particles become entangled, their identities merge: their collective state is precisely determined, but the individual identity of each of the particles has disappeared. The entangled particles behave as one, even when separated by a large distance."*
>
> *–Ronald Hanson, Professor at TU Delft's Kavli Institute of Nanoscience Netherlands*

If you are a spiritually attributed person, whatever or

whoever you are connected with in the spirit will overpower the physical. A spiritual connection is pointed to one's destiny, not to physical decay. But if the bond is carnal, ephemeral feelings will also die with the body. King Solomon said, there is no more work in the afterlife, so the one thing that you'll live up with is your memory, and you can't contest with that memory. The only thing that a soul can be proud of in the face of *cross-examination* is LOVE or he is being loved. Love saves us from all shame (See 1 Peter 4:8).

In the particle world, we wonder why matter is the only thing that appears on the scene. Antimatter is not very obvious in a state because as soon as it comes into contact with matter, it is annihilated and the entire mass of them both is converted into energy. From Einstein's E = mc2, where E stands for energy, m for mass and c for the speed of light, we know that energy and mass are related. From energy, mass is produced that generates equal amounts of matter and antimatter. When a photon splits in two and become entangled, the intensity they achieve will be alleviated in longer wavelengths equal to the energy of the original photon or source which was fired. It’s as if individually, they are empowered to merge in resemblance to their original form. It is a conservation of energy- in a higher degree.

The death of one twin is both an annihilation and an emergence of some information that will be retrieved as a sum, which will be “readable”. Most of his substantial information may be merged into his twin or soul mate's spirit who is physically living, and her features (memory) will also be glimpsed through him in eternity. The seed from his memory is replanted in her mind and it may serve in the physical plane through his soul mate’s “extended life”. This is “theoretically possible”, but not oftentimes existent since two worlds only correlate when their wavefunctions are non-zero (their value is positive and negative and not zero), that both must hold identical memories.

If a soul has a twin soul, she appends his information into her memory and her heart. As a representation to marriage, they are as two people countervailed as one in God’s scaling. Still, an emblem, a living “expressionism” that shows the sanctity of marriage. Twin souls are a catalyst of each other.

Is there marriage in heaven? We have to understand that marriage here on earth is a precursor of spiritual unions in the next life. It's a great blessing to find an earthly marriage that works as sanctification to reach your destiny or eternal reward. Twin soul is not a precursor of an earthly marriage, it is a precursor of a spiritual marriage brought forth to this earthly world.

"… And they twain shall be one flesh: so then they are no more twain, but one flesh." [xxix] Twain is an archaic term for "two", which is not only an allegory in wedding vows but a literal "lifestyle" for twin souls who long to justify the union but are yet being prepared in the "betrothal".

Jewish betrothal or *shiduchin* is a wedding engagement accounted as nominal matrimony solemnized long before the actual wedding. As the couple prepares themselves for the actual matrimony, the bride remains with her parents. The bride does not know when he will come back but she needs to be ready all the time. She keeps herself pure and dwells on him day and night as her love for him grows.

Beyond any secret, our memory is a storage that keeps countless bits of data information. Like Quantum Computer, 0 and 1 bits individually contain combined secreted texts. When you perform sex, there is incursion and transference of information between you and your partner. "Now Adam KNEW his wife, and she conceived..." (Gen. 4:1). [xxx] The keyword is KNOW, which alludes that through intercourse, the husband and wife will adapt each other's characteristics by spiritual identity.

Apostle Paul warned, "Don't you realize that if a man joins himself to a prostitute, he becomes one body with her? For the Scriptures say, 'The two are united into one" (1 Cor. 6:16). We wonder why a married individual who keeps an outside affair suddenly shifts personalities. Scientifically, promiscuity inveigles spiritual madness. (Check out a book, "Hooked: New Science on How Casual Sex is Affecting Our Children" by McIlhaney Jr. and Bush to learn more on this specific issue).

Two remarkable spiritual transferences in the Old Testament took place. "God took the spirit of Moses and transferred it into 70 elders of Israel, thereby reducing his responsibility and maximizing his purpose" (See Numbers 11:16-17, 24-25). Upon crossing the Jordan River, Elijah said to Elisha, "Tell me what I can do for you before I am taken away. Elisha asked, "Please, let me inherit a double share of your spirit and become your successor!" (2 Kings 2:9).

The spiritual link of Elijah and John the Baptist is truly mysterious and most particularly rare in terms of what kind of soul they are. Some say that Elijah reincarnated in the form of John but Elijah did not experience physical death. He was taken up to heaven body and soul (2 Kings 2). Based on Jesus' description of John in Matthew 11:11, John as a human, is another person. But in the spirit, John is Elijah, which Jesus continued on verse 14: "And if you are willing to accept what I say, he is Elijah..." They are like "soul twins" or part of one spiritual spectrum. In Luke 1:17, as Angel Gabriel announced that the old-aged Zechariah will have a son who is to be named John, the angel declared, "John will go before Jesus in the power and spirit of Elijah."

"It is absolutely required in the quantum teleportation particle that the thing that is teleported is destroyed in the process," said physicist Edward Farhi in an interview with physicist, Brian Greene. "That does make you a little anxious- but you would look too good." John Clauser laughed and said he wouldn't step in to that machine. [xxxi]

15
SYNCHRONICITY OR EINSTEIN RELATIVITY?

They say that twin souls are kept separated until the time is right. But what is the right time? And where? If twin souls will try to assess every event at a crossroad, the options will take them to no other possibility than whatever was determined. It's ironic to think that options can be made available- either to escape from the bond or to be a normal couple.

Lest you wonder where love hides in every pain, or why every relationship ends, perhaps your happy ending is not "right here", and forever is not here.

Around 1920s psychologist Carl Gustav Jung wrote to a Swiss author informing that Albert Einstein was his dinner guest and Einstein was developing the theory of Relativity. Jung believed that his meeting with Einstein was an opportunity for him to start a thesis on synchronicity. Jung, in his paper suggested that some coincidences are a product of thought or feeling or an event that coincide with another, somewhere, even in a dream or vision. He said, these are meaningful coincidences that cannot be explained by cause and effect.

"In a 1966 book, Thirty Years That Shook Physics, author George Gamow mentioned that physicist Wolfgang Pauli was associated with synchronicity events. "Pauli Effect" had become a terminology that refers to a mysteriously anecdotal failure of technical equipment whenever Sir Pauli walks in the lab. One early afternoon in Professor J. Franck's laboratory, a complicated apparatus for the study of atoms had collapsed. Franck humorously wrote Pauli, and Pauli replied right away saying, he had gone visiting Niels Bohr and at the time of the mishap in Franck's laboratory, his train was stopped for a few minutes at the Göttingen railroad station!" [xxxii]

Synchronicity is defined as experiencing two or more events as meaningfully related. "Carl Jung described synchronicity as "acausal connecting principle", "meaningful coincidence" or "acausal parallelism" that links mind and matter." [xxxiii] The number 11:11 is mostly believed to be a synchronicity number for twin souls. It is said that 11:11 also symbolizes rcflection.

The advent of synchronicity has brought about mystifications in the winch of signs and symbols. Alistine wrote, "Things like seeing numbers repeated everywhere, 11:11, having two complementary (or opposing) astrological signs, seeing your connection's initials all over the place, and so on. You're paying a lot of special attention to these details." [xxxiv] "Neurologist Peter Brugger said, a belief in interpreting some patterns and significance in license plate numbers, birth dates, and arrangements of fallen twigs, numerology, divination, prophecies of Nostradamus, etc., is called *Apophenia*." [xxxv]

The term "paranormal", if you look it up, is defined as something impossible to explain scientifically. For instance: miracles. Back in time, plagues, rare diseases, and pestilence were considered a raging blast from the gods. People are natural-born extremists, but today, devastations are clasped with increasing human activities and human interests for social media posts. The term, *Pareidolia* is the ability to perceive significant events through all circumstance without anticipating or calling out for signs and "lucky numbers". Sign is an event that travels faster than light because it occurs before the actual event.

At first, twin souls will think of synchronized events as tongue-in-cheek coincidences until these events become insurmountable. They have moved on in each of their own lives and made mistakes along the way, but synchronicity attests to them and shows them that they are a part of one course. Soon enough, they decide to choose each other until the end. Coincidences and synchronicity don't have to take place frequently, as long as it is the same combination and signs.

"Man, being the servant and interpreter of Nature, can do and understand so much and so much only as he has observed in fact or in thought of the course of nature."

–Francis Bacon (1561-1626), Philosopher

John Archibald Wheeler in 1940s, relating Quantum Electrodynamics said, "If one photon is vertically polarized, the other must be polarized horizontally. In order to effect such phase rotation, photons as well as electrons must be perfectly synchronized." [xxxvi] Wheeler was pertaining to the simultaneous rotations which particles take in opposite directions. In Albert Einstein's Relativity of Simultaneity, these particles inhabit two different worlds which events concur, directed by the velocity of light. "In 1900, Henri Poincare physically interpreted this Relativity as the result of Synchronization." [xxxvii] Lorentz Geometry shows that when two people or events are connected, they follow the Proper Time or Principle of Maximal Aging. If they run through a straight line, their life is harmonic and there is a chain of progression.

The number 11 that adheres to my twin soul life is usually associated with number 3. For instance, our first meeting is July 3 to 13 which makes 11 days. My business transactions as well as the publishing of this book start on our anniversary date.

Back in September 2, 1752, the date 3 through 13 that makes 11 days occurred as catastrophic event for seven million British people. They woke up the next day skipping 11 days of their life! September 3 through 13 were cut out from the British calendar losing 11 days! This calendric shift occurred because Pope Gregory XIII ordered a change from Julian to catch up on the Gregorian Calendar that they were compelled to cut down 11 days. September 3, 1752 was Thursday, and July 3, 2008 was Thursday, when Jason arrived in the Philippines (as if dragging 11 days from there). Similar shifts had also occurred in other countries in the ancient days but if some recent occurrences had dragged an entire calendar days from the past then it may be one incident of Lorenz's Butterfly Effect!

Calendars are simply human invention and so are clocks and watches but the ticking of time is said to be an account of gravity. As analogy and not a professed fact, let's say September 3-13, 1752 had left the British calendar space. Those days were only omitted from the calendar, but that doesn't mean they were also gone in space. The date might have caught up with space… but where will it settle? Will time by the speed of light rest on the same month in the future or on the same days? If it had made a stop in the Philippines time-frame on

September 3, 2008, Wednesday, the year 1752 is quite long only to get one day of inertial frame, hypothetically. But if the speed of light and gravitational mass are varied by days, 11 days to be exact, time might have compressed exactly 10 months where the difference in July 3, 2008 and September 3, 1572 is an account of "space-bent". Experts must take a look on this. It's a query, not a statement. If 11 days were gone, they will have found their own universal cycle somewhere, some time. We can trick aging but no one can trick time.

Time is not a physical system like how the classical concept perceives it by calendar days. That is why when people say, "forever", they define it based on the length of calendar days and how long the "waiting" could take. But Einstein said, time is not static, time goes with the speed of light and not by calendars. In Einstein's Special Relativity, all inertial frames are not equal, despite the constancy of light velocity. He said, Time cannot stand alone- it is melded with space by motion and direction. He thought that space and time adjust themselves in accordance to the speed of light and gravity which explains why our universe experiences time dilation and length contraction. If you turn left, your motion is shared by space on the other side. That's what "space is bent" means, as space is associated with time. Thus, the statement, Time is Spacetime.

Spacetime in our observable universe is made up of space with 3 dimensions- width, height, depth plus time as the 4th. In Einstein's General Relativity, time and space in a 4-D is continuum (that is why it's general relativity). This 4D spacetime is like a board that contains all events of my life from past to the future. So we don't imagine 4D as a span of one year- it is our whole lifetime which is likely to be dilated to 5-6D after this physical life. The song, "If Ever You're In My Arms Again" by Peabo Bryson describes "second chance" as a "second once in a lifetime," which could mean that "second chances" are not a matter of months or years but it must be a lifelong progression.

Relativity is reciprocal. His life and my life- the past, present, future, are set down in all Spacetime. Imagine two different worlds in opposite directions with synchronized events running off in the speed of light. If we are relative, the light relative to my distance and location and the light relative to his distance and location coexist in every angle. We are in a relative motion, but we appear different due to the time dilation. The basic postulate is that two observers in the same point of origin and inertial frame are as different observers, and yet are still synchronized. To other people's point of view, surely we travel at different distances and separate events, but in our point of view, we travel parallel to one another. This is because as observers, we share the same properties such as length, mass, time, that are relative to our motion, even though our clock differs. The mathematical logic of this is simple: If two events are "Mutually Exclusive", it is impossible for them to happen together. For example, head and tail, left and right. The time of two identical atomic clocks located at Colorado and England shows a matter of seconds' difference from each other. Colorado clock ticks a few micro-seconds ahead per year than the one in England does. No one can decide which one is accurate because both are accurate in their own way. The difference simply describes time in a matter of gravitational mass. In reference to Entanglement, Einstein proposed that what happens in one place does not always immediately affect the other. He was relating to how time bends with space (Spacetime Curvature). If you notice a pendulum swinging back and forth from point A to C, where C cannot meet with A, there comes a point that both will have to hold still so they can be relative to each other again. However, through the Lorentz transformations, differences can appear equal in which the totality of our events as two different sets will be submerged. They are to be compressed in contractions based on Lorentz-Fitzgerald, in which the continuum of different events corresponds even in different time frames. Our events related to each other are no longer ranked on the past, because all will be attributed to the future. You will find that past, present and future, are just one and the same. Sometimes we say "forever" if we catch a glimpse of a beautiful memory. Your "Now" holds forever- no past, no present, no future. You cannot bring back the past for the new past is illuminated and expanded as you get closer to time. You can only restore yourself with it in the now. "Forever" makes you feel there are no years passing by- every year is like every day.

Recently, physicists have discovered a jewel-like geometric structure called, *Amplituhedron*, introduced in 2013 by Nima Arkani-Hamed and Jaroslav Trnka. It is said that space-time merely arises as a consequence of this geometry, in which property is the adjoining of particle interactions. Since superposition is difficult to quantify, Amplituhedron allows us to calculate the particles' unitary interactions in the locality. Hence, if atomic clocks cannot both be the same, then the gravitational constant, even though constant also changes, because it adjusts to every atomic or quantized embodiment in our vacuum space. Red Shift shows that by a "regression curve", the speed of light changes at some period of time. It is said that time was faster in the days of Solomon and twice as fast in Abraham's.

If two people synchronize, life enfolds an order not only in our natural world but in God's time. Based upon a light-cone theory, an event can only influence another event when it is set inside the light cone. A light cone is a path of a single point in space and a single moment in time, directed by light that split as a hyperboloid. The original axis is called the Past. If an event is moved to the

Past, another region of event is influenced. Events within the light cone run from frame to frame as a series of parallel events. If Event 2 was caused by Event 1 in a certain reference frame, it will continue to flow. These events are like pearls lined up and held up by a string as though a pearl necklace. This necklace is called, *Worldline.* The speed of light is the one forming a worldline in the spacetime map. Worldline is represented by three types of spacetime intervals: time-like, light-like, and space-like intervals. Light-like interval is the most special because the events it enfolds can be causally connected through the speed of light. Both of your events remain relative despite the differences and changes.

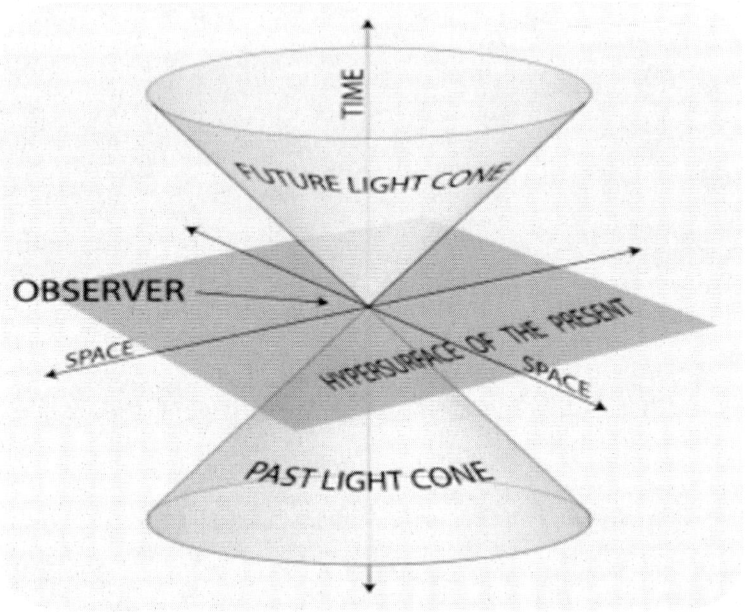

A time-like interval is the time between two events; you may also experience coincidences like seeing similar numbers or anything that reminds you but if some events are separated by space-like interval, then one of you is outside the light cone and you may not be causally connected. Ultimately you will know that you are connected when, even though he's gone, you are inspired to pursue his aspirations.

You can only imagine that the kind of life you're living now is actually a route you take to your next stop. They say it's destiny, and you design it your own and it's true, because once you're there, there's no turning back for you to say, I wish I took that road I once passed by.

We don't see our entire life, but God holds the Spacetime Map. You're a book in a vast library and God sees every single page of you- the past, the future, the whole piece. But sometimes we are given the privilege to taste our destiny for a moment so we can remain in a particular frame without getting lost or without losing each other. Whatever coordinate system twin souls are as two different observers, their lives are part of this invariant.

16

GUARDIAN'S KEEPERS
QUANTUM CRYPTOGRAPHY

When somebody breaks your heart, you simply forget and move on, because you made a common mistake by choice. But when you're hurt yet it has forever transformed you, you simply remember and gratify. The Lord knows that we forget easily, that we tend to repeat the same old mistakes. Deuteronomy 32 is a song which God asked Moses to compose and be taught to the people of Israel, to constantly remind them about God's grace bequeathed through every generation.

In Quantum Computing, "qubits are inherently faulty; they tend to forget some stored information. They find discomfort in unstable environments, and they are prone to flipping. Developed by Austin Fowler, these qubits find an ability to detect their own errors called, *Surface Code*- it allows the qubits to replenish reliability for superconducting quantum computers. Surface Code is safeguarding the appropriate data information as well as correction system that can store longer." [xxxviii]

The photon world is affected by the hazards we experience in this world; at times these photons break away or let go in exchange of strength and defense by giving up their comfort zone and allow “faith” to take over. Researchers developed a technique called, *Quantum Steering* that allows the photons to recover the security by getting reprogrammed with additional photon qubits. With these extra programming photons, any hacked measurement device can't extract the information encoded. Proposed by Seth Lloyd, there is *Quantum Illumination* that empowers particles to work in environments too noisy for entanglement.

Photon qubits are very safe and secure, as we know that these bits individually take 0 and 1 simultaneously which makes the system appear random. An eavesdropper wouldn’t get enough time to measure and “cryptographic erase” the key. The attempt will be known that a hacking is being performed and the twin photons will alert. Since they have opposite polarity, they will automatically oscillate in different directions. If someone tries to measure their state, we can detect the eavesdropping and so we reject the keys. If there is no hack, we apply the key to encode a message with absolute security. We repeat this on many entangled pairs that will form like a long string, and we can then generate a secret key.

Cryptography is used at protecting e-mail messages, credit card information, and corporate data. The art of protecting information by transforming it (encryption) into unreadable format is called *Cipher Text*. Data, often referred to as *plaintext*, is encrypted using an encryption algorithm and encryption key. This process generates cipher text that can only be viewed in its original form when it is decrypted with the correct key.

The word ENCRYPTION comes from the Greek word *kryptos*, meaning hidden or secret. The use of encryption is nearly as old as communication itself. As early as 1900 BC, an Egyptian scribe used non-standard hieroglyphs to hide the meaning of an inscription. Encryption schemes were made to convert messages into unreadable groups of figures to covert the messages while being transported elsewhere. Decryption is simply the inverse of encryption, following the same steps but reversing the order in which the keys are applied. Today's encryption algorithms are divided into two categories: symmetric and asymmetric. Only those who possess a secret key can decipher (or decrypt) the message into plaintext. Encrypted messages can sometimes be broken by Cryptanalysis, also called "codebreaking", although modern cryptography techniques are virtually unbreakable.

When a person gets "inspiration", anxiety mounts. This poor soul who cannot confide in a minister meets with a psychic. The spirits that guard the psychic mimic the information - twist it, and relay it to others. If you recall, Jesus told Peter, "Blessed are you because it is revealed to you by my Father in heaven". Then He declared, "You are Peter and on this rock, I will build my church, I will give you the keys to heaven". This also means that our key is very personal and we must treat a heavenly revelation exclusive to God's timing rather than spread it needlessly. Not everything you think that's guidance is guiding you. It could also confuse you or simply distract you from your important task.

I also experience a kind of test that only tests my character. Mostly it leads me to make a positive move, but by learning to move slow, I realized it wasn't God's will for me to really pursue it. I would know if it's His will when a guidance to do something persists. As long as the move is positive, I tend to do it, but if it does not result to another event, it could also mean that I should hold up. Qubits have preserved, original information that must remain unobserved. You are not allowed to know everything; just pull up enough information to detect errors, otherwise, the bit will drift again.

Matthew 16:19 says, "I will give you the keys of the Kingdom of Heaven; whatever you forbid on earth will be forbidden in heaven, and whatever you permit on earth will be permitted in heaven." Keys are for binding and loosing, opening and shutting, and people who believe in their hearts, "You are the Christ", as what Peter declared in his heart, are given the privilege to obtain this key.

Good and bad, wealth and poverty no longer seem to be the course of dispute these days. Anyone is attacked in all ways of cipher temptations and distress. Success, money, depressions, and relationships are forms of manipulation by a devil eavesdropper. But if we are in Christ, He will pave the imperfections or errors for us as we devote ourselves in the encryption and decryption of prayers. We also make it a commitment to protect the relationship and keep the love secure.

Essentially, Cryptography is based on the usage of photons. Other particles could be used, but photon, the particle of light, offers all the qualities. Their behaviors are comparatively well-understood and they are the information carriers in optical fiber cables, the most promising medium for extremely high-bandwidth communications.

By a certain polarization scheme that comes in a random sequence of bits, I send a key (0 or 1) to my partner (Encryption). This key is a stream of photons, each representing a single bit of data (0 or 1) interacting in several states of superposition. Let's say, "up/down, left/right, upleft/rightdown, and upright/leftdown". When photons are ready, they can pass through a polarizer. This time I can decide what to use whether rectilinear (up/down and left/right) or diagonal (upleft/rightdown and upright/leftdown) polarization filters. When receiving the photon key, my partner will also measure each photon bit using either his diagonal or rectilinear polarizer.

Sometimes he will choose the correct polarizer, other times the wrong one. We are not experts, but good partnerships can discard whatever wrong measurements chosen; and even if we use a correct polarizer, some errors can still happen. The outcome is the Post-Selection mentioned in Chapter 10, a process performed in Weak Measurement, in which after all the oscillations and symmetrizing through trials and divergences, the couple will be left with a number of states that make them "look good" together. A soul who has a twin is a person who is blessed with "somebody" who pleads his cause in the spiritual level from the physical level. This is a spiritual representation of Christ pleading for us from our physical level and putting a spiritual seal on us, calling us His own.

When the same mistakes repeated will finally go away and the lessons have rooted deep, you will realize that what kept you together was not the many second chances... the entire experience was in fact a "polarization scheme" for you to come out full, sieved, and not scattered. But the *sine qua non* is to be entangled with Christ. Without Christ in our life, we will be truly lost from this physical world to the next. Not all of us have a twin soul, and not all who claim as twin believe in Christ. Connection has a source and it's pointless without pointing that connection to the source it represents.

By a "double-channel polarizer", a soul can be determined and identified because he / she can be reflected through the image of Christ. No matter who we are, the scale which God uses to measure us is when He made us right in His sight, when we placed our faith in Christ (See Romans 3:22 & Romans 5:1-2).

> *"Well then, if we emphasize faith, does this mean that we can forget about the law? Of course not! In fact, only when we have faith do we truly fulfill the law."*
>
> *–Romans 3:31*

My favorite parable of Jesus is the "Parable of the Shrewd Manager" in Luke 16. The manager was shrewd and prideful. Remember that Jesus used parables to illustrate His Kingdom so we will understand. There are two kinds of people in this story: "of the world and of the light" and both are given the same privileges. The Master asked: "What is this that I heard that you wasted my goods? Do you know that you will be fired?" The manager was guilty and because he's arrogant enough to beg, he approached his coworkers and paid some of their debts so that if ever he gets fired, the workers will be kind enough to give him a place to stay in that Kingdom. The basic points of this story are: the manager was wise enough to recognize that worldly possessions are useless compared to the life he has in his Master's kingdom (see v. 9) and the voice of the children of light matters to the Master. I am a part of that Kingdom, as a lot of us are, and I am also a manager in one of the Master's estates. In the next life, we will give an account. Based on v. 8, I'm hoping that the Master will say, "Well then, I never thought that in this generation, there is a child of light that is more "shrewd" than that of the world. Very well, whatever you ask for this man will be granted."

A more advanced technology than Cryptography gives us a reflective thought how special we are in every bit. It's called QKD (Quantum Key Distributions). In terms of encoding information, QKD doesn't use polarization- it transfers bits per photon by its orbital angular momentum and position (OAMP). This was first conducted by the researchers at the University of Rochester. Scientists otherwise call OAMP a "twisted light". It empowers 1 bit to increase the efficiency of cryptosystem. It shows that light can be released and benefited besides the selection filters of polarization. If two "managers" complement in terms of OAMP, they are able to encode a seven-dimensional "alphabet" which letters produce a secure key.

> *"Unlike with polarization, where it is impossible to encode more than one bit per photon, "twisted light" could make it possible to encode several bits, and every extra bit of information encoded in a photon means fewer photons to generate and measure."*
>
> *–Mohammad Mirhosseini, University of Rochester*

Looks like love will only get a happy ending when it is fully secured! The reason why scientists use entangled photons is because their bond is just too strong to be mimicked by hacked devices. Today, commercial prototypes of simple QKD systems have been used to protect information transfers between bank branches as well as quantum-enabled smart phones to transact with an ATM.

THE CROSSROAD

"I'm nervous about the whole thing, but my faith, trust and respect for you give me peace. I think my loneliness and lack of any real responsibilities have the majority to do with those shortcomings throughout my 20's. I don't have anyone to be accountable to, you know? I hope that stuff doesn't worry you though. I feel that I need you in my life. I have zero hope that I would ever meet someone who I could trust as much as I trust you. I suppose the point of a relationship is to take on life as a team and make each other feel secure and loved. I've dated enough women in my day to no longer be interested in "testing the waters", so to speak. I would take pride in being faithful to my baby."

–Jason, July 23, 2013

How a wound heals is how a book closes. Healing brings clarity to the mind, to the premise. He left knowing what happiness is and knowing what it's like to live without it. Sometimes it's better not to find what makes you happy…

but not knowing it is not finding your destiny either. Everyone who knows my beloved, may think that he lived a short, empty life. He was a regular guy to others, but that view of him wasn't his real world. He changed my life- that was the real him, and the decision to give my life to Christ was like putting a seal to the relationship. It was a strong conviction that I decided to be re-baptized in a small Baptist church (March 1, 2014). With Jason, it was a real Christian experience which I never had when I was growing up, even when I became active in church activities. When I let him go, I begged God to keep him safe, no matter what... a spiritual prayer which begot more answers too wonderful to describe. The start of our last comeback was May 2013, the same month we first met (The last conversation is off the record).

Jason died on July 2014. It was his birthday when I learned of his death, 2016, and I was close to finishing this book. For the first time I sent a message to any of his family, to his brother, Josh, who passed away one and a half month after we spoke. I spoke to Jason's close friend, Joseph, and step-brother, Brock and I'm thankful to them.

My last encounter with him, October 2013, was a “rebirth”, or a spiritual repolarization. I decided to surrender my life to God, including his as everyday I tried to hold on. Knowing from my experience as a fake Christian before, I had quite a struggle on my path towards light. I was worried whether Jason could make it either. But I had to let go. I was deleting my email account, but Yahoo was down then I lost internet for a week. Suddenly I got a foreboding sense that something “bad” might happen so I emailed him instead. Since there was no reply, I stopped emailing as I could also sense a new condition. It was peace and great love. My worries went away and I felt strongly motivated to finish this book. When I learned about his death, I recalled everything, I was shown the entire picture and I reviewed what I had accomplished. The connection still feels the same except that there is no more changing weather, and so I know that he is in a better place. Two hearts colliding... it’s a world that's as close as what I am.

“It’s hard to describe but being in Negros is like being born again, in a world where I am not so self-conscious, not so materialistic, and also being one with the land," Jason uttered, May 2013. "No police record, no past mistakes, like I said, like being born again but with the wisdom of my rocky life before."

I was happy and excited checking out scooters at a shop, imagining the best one for him as he asked me to. He was planning to pursue his schooling in Negros, and so I visited two prospective universities. I know for a fact that going back to Negros was his dream, he just had to be sure and ready for it.

"I do not say it because this is not the type of relationship we have now but I LOVE YOU. Let that be understood." Jason's mail surprised me that June 24, 2013. For the first time after five years I truly felt what I am as a person, what I certainly am in Jason's life- that I am his other half.

For three days, I had fears running through my head, hoping that Jason was serious about his plans, but more than anything, I understood how huge of a turning point it would be for him. He said he would be joining me in December 2013. What really worried me was whether I should wait some more if ever he wouldn't make it. I realized then that life can be scary. I knew I could just say 'No' and walk away whenever I found something to be unbearable, but not with Jason, and that's what scared me. I tearfully called out to God for help and for the first time I uttered, "Angels, please, whoever you are, help!"

"Hey sweetie, I am so glad I had the opportunity to check my mail and you're the only person I care to hear from… I miss you. I look forward to seeing you again. Oh, also, I love my baby. I speak very highly of you to people around here. I'm just so proud of you." I heard from him the following day which was his 30th birthday.

For the first time, I started an adventure from the crossroad going to Quezon. It is a peaceful little village where ricefields and hills are simply heavenly. Three kilometers away is Don Salvador which is now the Summer Capital of Negros Island. Three days later, I received an email from Jason talking about crossroads.

"The first path would be the wrong one- fast life, drugs, partying," Jason shared about the Crossroads he learned from his meditation. "The second path is a road that seems empty but somehow you make progress- me being in college and moving forward but certainly not happy. The third path is the pursuit of true happiness. I think you and Negros might be my actual path to happiness."

All of the sudden, you pushed me away as far as you could sprawl. We were happy, we had been more than friends for many years, we shared our dreams together and we had future plans as well as plans to see Batanes. You knew I'd be worried. I knocked and totally you ignored me, shut me off. Could it be that easy to throw away the most splendid thing one can ever have? Then I let you go so you could live a normal life, marry somebody else or just live a life without me. I stopped thinking that you love me and I stopped believing that you care about me. I was hoping that you would tell me that I don't matter. It could be easier to live that way, waking up to a reality without you. But you drag me everywhere and make me feel what you feel. It's severe to realize about you, that the way you mean to me is the way I mean to you.

Today is January 13, 2017.

Is it over for Jason and Stella?

To find out how a *ghostly* love can't get enough, grab a copy of…

My Husband in Heaven

This book offers an advanced Physics to illustrate the love that overpowers death featuring:

Cavity Quantum Electrodynamics

Neurophysiology

Thank you for reading.

ENDNOTES

[i] From Aristotle's major treatise, *On the Soul* 412b5-7, 413a1-3, 414a15-18

[ii] Frigg, Roman and Hartmann, Stephan, "Models in Science", The Stanford Encyclopedia of Philosophy (Spring 2017 Edition), Edward N. Zalta (ed.), https://plato.stanford.edu/archives/spr2017/entries/models-science/

[iii]. Rabbi Berachya and Rabbi Chalbo and Rabbi Samuel b. Nahman's commentary on Deuteronomy 4:32, Genesis 2:21, Psalm 139:5 and Midrash Yalkut Shim'oni's commentary to Exodus 26:20.

[iv] The primary idea of a Jewish soul mate comes from the statement by R. Yehuda (Rav) in Sotah 2a.

[v] Degrees of Freedom, https://en.wikipedia.org/wiki/Degrees_of_freedom_(physics_and_chemistry)

[vi] Virtual Particles, Superstrings, P-Branes and M-Theory (PediaPress), 181.

[vii] Based on Hylomorphism by Aristotle

[viii] Sota 9:3 (45b)

[ix] 1 Corinthians 15:45-46- New American Standard Bible

[x] Young Two-Slit Experiment. http://abyss.uoregon.edu/~js/21st_century_science/lectures/lec13.html

[xi] Lubos Motl, The Holographic Principle, 26 October, 2012, http://motls.blogspot.com/2012/10/the-holographic-principle.html

[xii] Martin Chaplin, "Icosahedral Water Cluster Architecture", 21 July, 2015 http://www1.lsbu.ac.uk/water/icosahedra.html

[xiii] Tibi Puiu, A Checmical Bond is Born, 13 February, 2015 http://www.zmescience.com/science/chemistry/chemical-bond-image-first-time5345

[xiv] NASA/WMAP Science Team, Foundations of Big Bang Cosmology: https://map.gsfc.nasa.gov/site/citations.html

[xv] Ibid.

[xvi] In 1884, the French chemist and engineer Henry-Louis Le Chatelier (Le Chatelier's Principle), describes what happens to a system when something momentarily takes it away from equilibrium.

[xvii] https://en.wikipedia.org/wiki/Quantum_electrodynamics

[xviii] Physical Constant, https://en.wikipedia.org/wiki/Physical_constant

[xix] Sofia Siberia, "The Merging: Twin Flames", http://sofiasiberia.com/2013/12/23/twin-flames/

[xx] Original item was written by Erin Bow reprinted from materials provided by Perimeter Institute for Theoretical Physics: Are weak values quantum?, http://www.sciencedaily.com/releases/2014/09/140924112826.htm

[xxi] Kadari, Tamar. "Eve: Midrash and Aggadah." Jewish Women: A Comprehensive Historical Encyclopedia. 20 March 2009. Jewish Women's Archive. (Viewed on December 29, 2016) https://jwa.org/encyclopedia/article/eve-midrash-and-aggadah

[xxii] Amir D. Aczel, Entanglement: The Greatest Mystery in Physics, (Four Walls Eight Windows New York 2001),p. 25

[xxiii] John Byl, Indeterminacy, Divine Action & Human Freedom, http://reformation.edu/scripture-science-byl/pages/07-indeterminacy.htm "John Byl is professor of mathematics at Trinity Western University."

[xxiv] Originally from Healing Codes for the Biological Apocalypse pages 166 & 167 by Dr. Leonard Horowitz and Dr. Joseph Puleo: http://www.schumannresonator.com

[xxv] http://www.schumannresonator.com/

[xxvi] https://en.wikipedia.org/wiki/Lamb_shift

[xxvii] Chemistry LibreTexts: University of California, Zeeman Effect in Hydrogen Atom; http://chem.libretexts.org/Core/Physical_and_Theoretical_Chemistry/Quantum_Mechanics/Zeeman_Effect/5%3A_Zeeman_Effect_in_Hydrogen_atom

[xxviii] Columbia University- Department of Physics, "Quantum Criticality", http://www.physics.rutgers.edu/qcritical/frontier3.htm

[xxix] Mark 10:8, KJV

[xxx] English Standard Version

[xxxi] Based on a video from YouTube, "The Fabric of the Cosmos: Quantum Leap" by Physicist Brian Greene

[xxxii] George Gamow, Thirty Years That Shook Physics – The Story of Quantum Theory, (Doubleday & Co. Inc. New York, 1966), 64.

[xxxiii] Synchronicity, http://en.wikipedia.org/wiki/Synchronicity

[xxxiv] Alistine, "Twin Flame Signs and Syncs", http://aelestrid.net

[xxxv] Peter Brugger, "From Haunted Brain to Haunted Science: A Cognitive Neuroscience View of Paranormal and Pseudoscientific Thought," Hauntings and Poltergeists: Multidisciplinary Perspectives, edited by J. Houran and R. Lange (North Carolina: McFarland & Company, Inc. Publishers, 2001).

[xxxvi] Amir D. Aczel, Entanglement: The Greatest Mystery in Physics, (Four Walls Eight Windows New York 2001), 87

[xxxvii] Relativity of Simultaneity; https://en.wikipedia.org/wiki/Relativity_of_simultaneity

[xxxviii] Ausin Fowler et.al. Surface codes: Towards practical large-scale quantum computation: Phys. Rev. A 86, 032324 (2012); arXiv:1208.0928v2, DOI: 10.1103/PhysRevA.86.032324

Made in United States
North Haven, CT
04 November 2026

21770341R00116